INTERNATIONAL DEVELOPMENT IN FOCUS

Agricultural Innovation in Developing East Asia

Productivity, Safety, and Sustainability

RIIKKA RAJALAHTI

WORLD BANK GROUP

1818 H Street NW, Washington, DC 20433
Telephone: 202-473-1000; Internet: www.worldbank.org

1 2 3 4 24 23 22 21

ISBN: 978-1-4648-1681-9
DOI: 10.1596/978-1-4648-1681-9

Cover image: © monopoly919 / shutterstock.com. Used with permission of monopoly919 / shutterstock.com; further permission required for reuse.
Cover design: Debra Naylor / Naylor Design Inc.

Contents

Boxes

Figures

Tables

Acknowledgments

This report was written by Riikka Rajalahti, a consultant and former World Bank Senior Agriculture Specialist on Agricultural Innovation Systems. The document was prepared as a background paper for the East Asia and Pacific (EAP) regional study *The Innovation Imperative for Developing East Asia*. The report was prepared under the overall supervision of Andrew Mason, lead economist; and Aaditya Mattoo, chief economist of the EAP Region; in collaboration with Paavo Eliste, lead agricultural economist, EAP; and Dina Umali-Deininger, practice manager, Agriculture and Food Global Practice, EAP.

Helpful comments were provided by Anwar Aridi, senior private sector specialist, Finance, Competitiveness and Innovation Global Practice, Europe and Central Asia Region; Indira Ekanayake, senior agriculture specialist, EAP; Ann Jeannette Glauber, practice manager, Environment, EAP; Willem Janssen, lead agriculture economist, South Asia Region; David Kaczan, economist, EAP; and Stephen Ling, lead environmental specialist, EAP. Eija Pehu, consultant and former World Bank science advisor, made contributions during the preparation of this report, which are gratefully acknowledged.

Executive Summary

AGRICULTURAL INNOVATION HAS PLAYED A KEY ROLE IN EAST ASIA'S DEVELOPMENT

Agricultural innovation has played a critical role in the economic transformation of developing East Asian countries over the past half century. This transformation began with the diffusion and adoption of high-yielding seed varieties, modern fertilizers, and other agricultural technologies (for example, pesticides, machinery), commonly known as the Green Revolution. A strong focus on increased production and food security, and rapid adoption of modern rice and corn varieties, coupled with investments in irrigation, agricultural extension services, and broad economic reforms, resulted in dramatic increases in crop yields and agricultural productivity across the region (figure ES.1). This higher productivity, in turn, contributed to higher farm incomes and lower poverty in rural areas. Higher agricultural productivity freed rural labor to move into manufacturing and services, and it enabled significant expansion and intensification of export-oriented commodity production. In short, innovations in agriculture helped lay the foundation for the structural transformation that has fueled subsequent growth and development across the region.

Despite significant structural changes in the region's economies, with rapid growth in manufacturing and services, the agri-food sector remains highly important for developing East Asia. The food value chain accounts for roughly 14 percent of the region's gross domestic product (GDP) and 35 percent of the labor force, with considerably higher shares in several lower-middle-income countries (figures ES.2 and ES.3). Nearly 40 percent of Vietnam's labor force still works in agriculture, for example, as does 50 percent of the labor force in Myanmar, and more than 33 percent of the labor force in the Lao People's Democratic Republic. Moreover, the region dominates global trade in tropical fruits and is gradually expanding exports in horticulture, seafood, and processed products. Agricultural sector performance thus remains important to the region's development, including for raising rural households' incomes, further reducing poverty, and meeting countries' growing food demand and nutrition requirements.

FIGURE ES.1
Increases in rice and maize yields in developing East Asia, 1961–2018

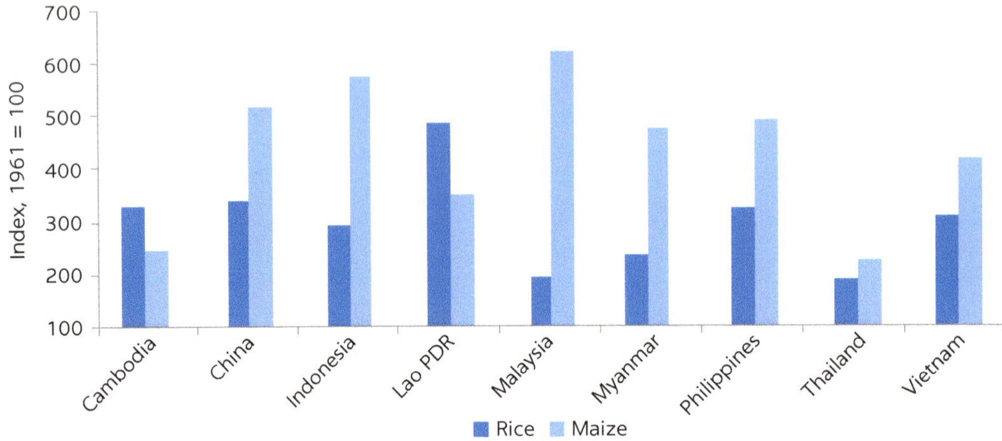

Source: FAOSTAT (Statistics Division, Food and Agriculture Organization of the United Nations), 2020 (http://www.fao.org/faostat/en/#data).

FIGURE ES.2
Agricultural value added, 2018

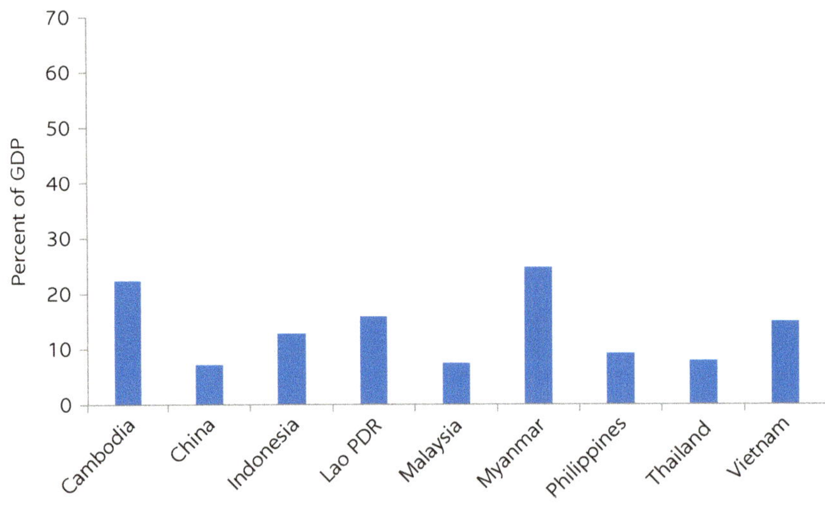

Source: World Bank Databank, 2020 (https://data.worldbank.org/).

FIGURE ES.3
Agriculture sector employment, 2018

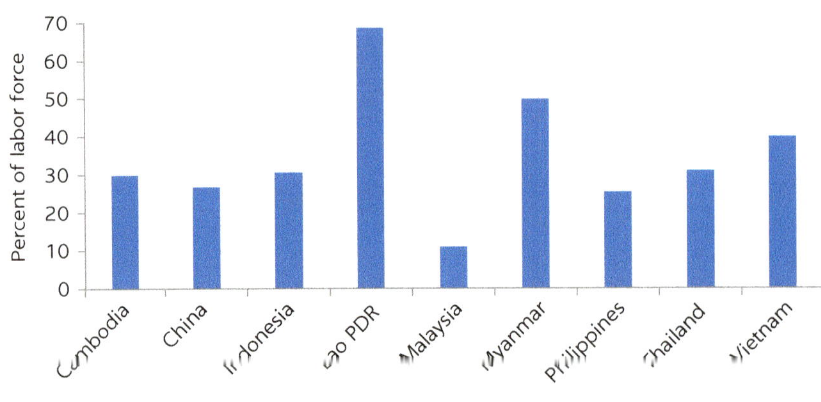

Source: World Bank Databank, 2020 (https://data.worldbank.org/).

THE REGION'S AGRICULTURE AND FOOD SYSTEMS ARE UNDER PRESSURE

With economic change, the region's agri-food systems are under increasing pressure to meet the demand for food. Domestic dietary and food expenditure patterns are changing as incomes rise and societies become increasingly urbanized. Consumer demand for staples such as rice has declined whereas consumption of more nutritious, higher-value fresh food and animal products and packaged convenience foods has increased. However, small and fragmented farm structures and aging farmers, often with relatively low education levels, constrain the sector's ability to invest in and adopt innovations for improved productivity and sustainability. Moreover, climate change has created new pressures on productivity, with key cereal crop yields in most of the region projected to decline. Supply-side challenges are further compounded by limited land and water availability and high levels of food loss and waste.

Agri-food sectors in the region are experiencing environmental stress and threats to sustainability. A long-standing emphasis on raising production, coupled with constraints on land and water, has had negative effects on the environment. Conversion of more marginal, virgin land for agriculture and overuse of water, agrochemicals, and veterinary drugs, combined with generally poor farming practices, have all had adverse environmental impacts. The effects are vast, including deforestation; loss of biodiversity; soil erosion; pollution of water, air, and oceans (plastics); accelerated greenhouse gas emissions; and water scarcity and salinization (table ES.1). Environmental degradation is negatively affecting agricultural productivity in many places. The expansion and intensification of agriculture has also resulted in habitat degradation and fragmentation, which has amplified the incidence of emerging infectious diseases and zoonoses originating in livestock and wildlife.

TABLE ES.1 **Principal environmental risks, by commodity, in East and Southeast Asia**

COMMODITY	GEOGRAPHIC AREA	SOIL DEGRADATION	POLLUTION	WATER SCARCITY OR SALINIZATION	BIODIVERSITY AND HABITAT LOSS[a]	GHG EMISSIONS[b]
Coffee	Tropical uplands	Dark blue	Yellow	Dark blue	Light blue	Yellow
Cocoa	Tropical uplands	Light blue	Dark blue	Yellow	Light blue	Yellow
Rubber	Tropical uplands	Yellow	Light blue	Yellow	Dark blue	Yellow
Bananas	Tropical uplands	Dark blue	Dark blue	Light blue	Yellow	Yellow
Palm oil	Lowlands and midlands	Dark blue	Light blue	Yellow	Dark blue	Dark blue
Tea	Tropical uplands	Light blue	Dark blue	Yellow	Dark blue	Yellow
Maize	Varied	Dark blue	Dark blue	Dark blue	Light blue	Light blue
Rice	Varied	Dark blue	Dark blue	Dark blue	Dark blue	Dark blue
Sugarcane	Lowlands	Dark blue	Dark blue	Dark blue	Yellow	Dark blue
Pork (intensive)	Varied	Yellow	Dark blue	Yellow	Yellow	Dark blue
Shrimp (intensive)	Coastal lowlands	Yellow	Dark blue	Yellow	Dark blue	Yellow

Source: Scherr et al. 2015.
Note: Dark blue = high risk; light blue = medium risk; yellow = low risk.
a. Including forest clearing.
b. GHG (greenhouse gas) emissions from land uses and land clearing.

The costs associated with environmental degradation are substantial. At the individual farm and community level, costs are associated with decreasing productivity and lower production, along with a wide range of health effects. However, environmental costs associated with agriculture go well beyond farms. In the East Asia and Pacific Region, the annual cost of ocean pollution from plastics, some of which originate from the agri-food sector, is estimated at more than $1.3 billion. Zoonoses, such as COVID-19 (coronavirus), can have devastating effects on human capital through loss of life and livelihoods, greater malnutrition, and the interruption of essential health and education services. The costs of managing zoonotic outbreaks on an annual basis are estimated at more than $30 billion, not considering the costs associated with the COVID-19 pandemic. Indeed, estimated economic losses associated with environmental degradation, as captured by natural resources depletion, are significant across developing East Asian countries (figure ES.4).

Food safety problems have created new food insecurity as the region rapidly urbanizes. The available evidence also points to the rising exposure of populations to food safety hazards, a significant incidence of foodborne illness (figure ES.5), and deepening consumer concerns about the contamination of local foods and the adequacy of prevailing governance structures to manage emerging risks. The region's food safety problems stem from a variety of sources across the agri-food system, from the farm, to food processing and retail, to households. These problems have been compounded by the rapidly increasing consumer demand for purchased and processed foods and the inability of countries' food safety systems to keep up with changing circumstances. When considering the productivity losses and cost of treating foodborne diseases, the annual domestic costs of unsafe food to human health and to the economies in developing East Asia amount to more than $45 billion (figure ES.6). A crude estimate of the trade-related costs due to food safety violations is equivalent to roughly 3 percent of the value of high-value food exports across emerging Asia.

FIGURE ES.4

Cost of natural resources depletion in developing East Asia

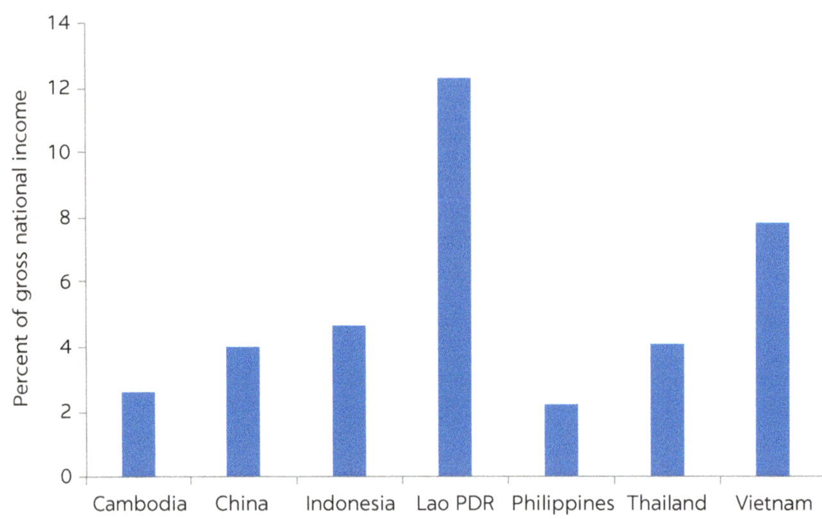

Source: Adapted from Scherr et al. 2015.
Note: Natural resource depletion (2012 data) is the sum of net forest depletion, energy depletion, and mineral depletion. Estimates exclude the economic costs of greenhouse gas damage and ecosystem service losses.

FIGURE ES.5

Foodborne disease in a comparative perspective, 2010

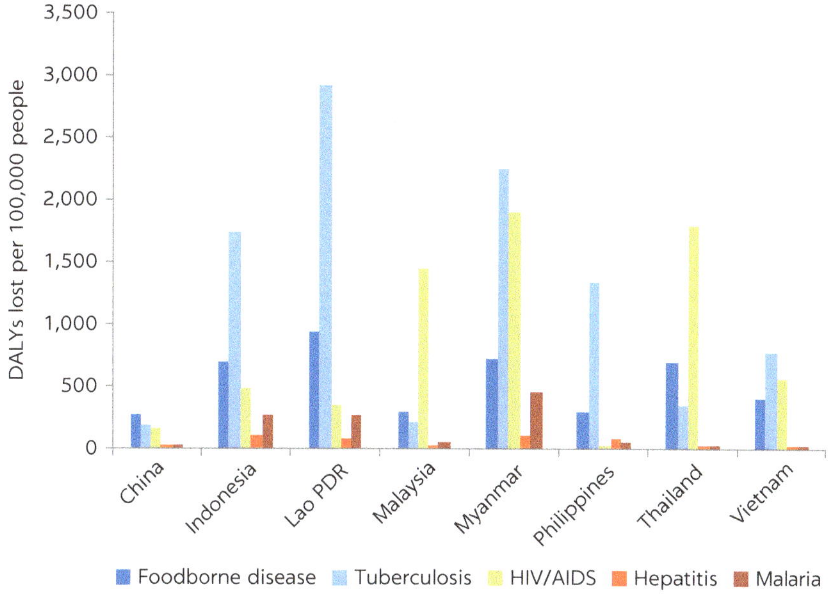

Source: World Health Organization Global Burden of Disease Statistics and Foodborne Disease Epidemiology Reference Group; Ekanayake, Ambrosio-Albala, and Jaffee 2019; World Bank 2019.

Note: DALY = disability-adjusted life year; HIV/AIDS = human immunodeficiency virus /acquired immune deficiency syndrome.

FIGURE ES.6

Estimated productivity loss due to foodborne disease, 2016

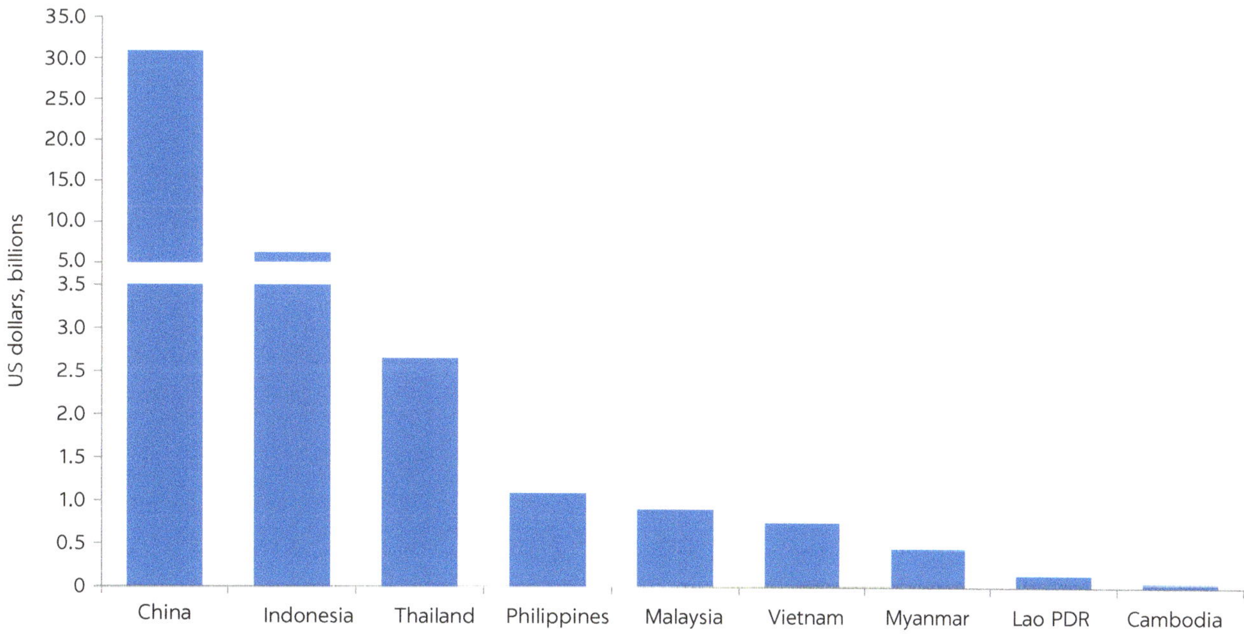

Source: Adapted from World Bank 2019.

Food safety concerns are compounded by dietary quality challenges, such as imbalanced diets, micronutrient deficiencies, and the rising incidence of overweight and obesity. The region's success in addressing this "triple burden of malnutrition" will have strong implications for progress on human development over the coming decades. Investments in human capital are becoming more important as the nature of work evolves. By improving public health, in addition to skills and knowledge, people can be more productive and innovative.

The COVID-19 crisis has also exposed long-standing and emerging weaknesses in the region's agri-food systems. The systems have come under growing pressure to address the economic, environmental, and health-related weaknesses associated with expanding and intensifying production. The COVID-19 crisis, with its widespread effects on livelihoods, food security, trade, and health, is a hard reminder of the prevailing demand- and supply-side constraints affecting the region's food systems. Although food supply chain disruptions and resultant export bans were only temporary, the pandemic has revealed deeper structural problems—such as disruptions caused by the concentration of supplier bases, limited inclusion of farm populations in public safety nets, and unhealthy diets that have raised obesity and other noncommunicable diseases and, in turn, worsened COVID-19-related health outcomes—that are expected to last well beyond the crisis. In this context, the COVID-19 crisis could also offer a window of opportunity for governments in the region to prioritize increasing the resilience of agri-food systems as part of rebuilding more sustainable systems, which would also better address public health concerns. The key task is to transform profound crisis-induced changes into the foundation for improved agri-food systems. Innovation can play a critical role in this transformation.

This report examines the state of agricultural innovation in developing East Asia. It assesses countries' preparedness to undertake innovations to address the region's triple challenge: increasing productivity, improving food safety, and enhancing sustainability. And it lays out an agenda for agricultural innovation systems that can foster transformation toward delivery of a "triple win." The report is based on a review of existing data and literature with a focus on nine developing East Asian countries: Cambodia, China, Indonesia, Lao PDR, Malaysia, Myanmar, the Philippines, Thailand, and Vietnam. All nine countries are considered middle-income countries in the World Bank's classification system, but there is considerable diversity in country circumstances, including in innovation capabilities. The intent of the report is to identify key constraints for innovation in agriculture and to highlight key areas for policy action. Given the substantial heterogeneity across countries, deeper country-level analysis would be needed to prepare more detailed country-specific recommendations.

POLICY MAKERS RECOGNIZE KEY CHALLENGES FACING EAST ASIA'S AGRI-FOOD SYSTEMS, BUT MORE ACTION IS NEEDED

Policy makers have increasingly realized that the same farming techniques that have raised agricultural productivity are undermining long-term agricultural sustainability, and that better approaches are needed. In response, starting in the 1990s, a number of developing East Asian countries began experimenting with more sustainable cropping and livestock practices. In southern China, for example, sustainable rice intensification, which reduces water and input use while increasing yields, is becoming a widespread practice. Integrated soil-crop management practices in rice production in China and Vietnam have reduced reliance on purchased inputs and contributed to greater environmental sustainability and health through improved use of agrochemicals, often with only limited productivity trade-offs. Despite their potential benefits, these types of innovations tend to be more knowledge and management intensive, limiting their adoption in the countries where agricultural production systems remain dominated by smallholders.

However, these innovations have often been undermined by more traditional support programs, which have distorted farmers' incentives to make appropriate decisions toward achieving the triple wins. For example, a significant share of public agriculture spending has gone toward financing input subsidies and price support programs, mainly for cereals, contributing to low adoption of more sustainable farming practices. At the same time, important public goods that would benefit farmers and society, such as extension and environmental services, are undersupplied, reflecting the tight fiscal space of many governments and continuing political support for distortionary input subsidies. This bias explains the relatively limited adoption in most countries of improved, knowledge-intensive farming practices, whether integrated soil-crop management, sustainable rice intensification, or climate-smart agriculture.

Although government initiatives to increase farmers' capabilities and improve incentives have in some instances paid off, more—and more systematic—action is needed. Evidence suggests that the impact of farmer education on farm productivity and income is greater in more economically and technologically dynamic environments, as well as where policy makers have put in place measures to support the adoption of more environmentally sustainable technologies and practices (for example, slow-release fertilizers, soil testing kits, and so on). Improved and relatively easy-to-adopt, cost-effective technologies, such as biotechnology and information and communication technologies (ICTs), have also supported smallholders' ability to increase productivity in a more sustainable manner. However, numerous opportunities remain to reform existing agri-food policies and support programs, such as distortive input subsidies and commodity price supports, to encourage wider adoption of sustainable technologies and enable farmers and firms to address the growing challenges facing the region's agri-food systems.

A NEW GENERATION OF INNOVATIONS CAN TRANSFORM THE REGION'S AGRI-FOOD SYSTEMS, INCREASING PRODUCTIVITY, SAFETY, AND SUSTAINABILITY

Agricultural innovations have enjoyed high returns in the region. Indeed, agricultural research and development (R&D)—particularly to support the adaptation of new technologies and practices to local contexts—has been important to increasing agricultural productivity in the region, generating returns of nearly 50 percent to public spending on R&D in agriculture. Rapid changes in the region's economies are leading to the need to rethink countries' approaches to agri-food system development, however, looking beyond productivity to place greater emphasis on food safety, resilience, and agri-food sector sustainability.

Harnessing a new generation of agricultural innovation will be critical to transforming agri-food systems in developing East Asia. Several transformative innovations hold great potential to increase agricultural productivity, food safety, and agri-food system sustainability, and in doing so build greater resilience to external shocks. Many of these transformative innovations already exist, and the primary focus needs to be on adaptation to local contexts and, subsequently, adoption and diffusion of new technologies and practices. In other cases, further scientific research will be critical, for example with respect to biotechnologies, to improve the resilience of crops to rising climate risk. Important advances have been made along multiple dimensions of agro-food systems, such as the following:

- Innovations supporting *agricultural production* that have already demonstrated great potential not only to improve productivity but also to increase agri-food systems' sustainability and resilience to external shocks
- Innovations along the *food chain* that have the potential to increase efficiency and food safety and improve health outcomes, through improved traceability, reduced food loss and waste, and lower greenhouse gas emissions
- Innovations related to *food consumption and nutrition*, including alternative food sources that can improve health outcomes and, in some cases, enhance environmental sustainability and efficiency

Although increased productivity remains an important goal in developing East Asia, many innovations show great potential to deliver a "triple win," that is, to enhance productivity, increase food safety and nutrition, and improve environmental sustainability (table ES.2). However, the feasibility of adopting these transformative innovations is highly context specific, both across and within countries. The ability of farmers to adopt potentially transformative innovations is influenced by the policy incentives that affect the cost of technology, access to finance, land ownership, ICT connectivity, and the level of farmers' knowledge and skills. In many contexts, for example, access to finance and land ownership remain barriers to smallholder adoption of precision agriculture or mechanization, which would require some form of land consolidation. Similarly, limited digital-mobile access and affordability may prevent some farmers from adopting ICT-based solutions for extension, weather, insurance, and logistics services as well as from pursuing precision agriculture and emerging e-commerce opportunities.

Developing East Asian countries exhibit different levels of preparedness to adopt or create transformative innovations. The ability to adapt, utilize, or create new transformative innovations varies by countries' level of science, technology, and innovation (STI) capacity as well as by the ease with which the innovations can be adopted in a given agri-food system context. Specifically, farmer-enterprise capacity, access to innovation and finance, existing infrastructure, and so on must be considered. Panel a of figure ES.7 presents an illustrative and subjective ranking of a range of innovations and the associated STI capacity requirements and ease of adoption. For instance, easy-to-adopt innovations include biofortification, improved varieties, and high-efficiency and less toxic agrochemicals, whereas greater skills and effort are required to generate and adopt many forms of biotechnology, precision agriculture, food safety applications, and urban farming innovations.

Countries' different levels of preparedness to innovate generally correspond to their overall levels of development (figure ES.7, panel b). The front-runners in nearly all aspects of transformative innovation are the upper-middle-income countries in the region—China, Malaysia, and Thailand—which have relatively high STI and adoption capacity. Several large lower-middle-income countries, including Indonesia, the Philippines, and Vietnam, are well positioned to develop or adopt traditional and new innovations (for example, e-services, climate-smart agriculture, genetically modified [GM] crops), although their capacity to address

TABLE ES.2 **Transformative innovations in agriculture and their potential to address productivity, sustainability, and food safety challenges in developing East Asia**

INNOVATION AREAS	PRODUCTIVITY	FOOD SAFETY AND NUTRITION	ENVIRONMENTAL SUSTAINABILITY
Production			
Precision agriculture	High	High	High
ICT-based solutions for extension, weather, insurance, logistics services	High	Low	High
Urban farming (vertical farming, hydroponics, aeroponics, growth kits) and protected farming	High	Moderate	High
Biotechnology: genetic modification, genome editing microbiomes, biopesticides, feed additives	High	High	High
Biofortified crops	High	High	Moderate
Sustainable agriculture practices (climate-smart agriculture; management of pests, soil fertility, water, land-use intensity; livestock production; organic farming)	High	High	High
Mechanization (farm tools and machinery, irrigation, agro-services, 3D printing of tools)	High	Low to moderate	Moderate
Food chain			
Traceability (blockchain, antifraud tools, sensors)	Moderate	High	Moderate
Food safety (diagnostic tests, antibacterial proteins)	n.a.	High	n.a.
Food loss and waste (data systems, e-commerce)	Moderate	Low	Moderate
Food loss and waste (storage solutions, smart labels, biodegradable packages, circular economy)	Moderate	Moderate	Moderate
Food consumption and nutrition			
Fortification, reformulation, functional foods, nutrigenetics	n.a.	High	n.a.
Lab-grown meat and other new food sources	Moderate	Moderate	Moderate
Other			
Organization and collective action-related innovations (extension, value chain, water user associations)	High	Moderate	High

Source: Original table for this publication.
Note: High = high potential to address the challenge; Moderate = some potential to address the challenge; Low = low potential to address the challenge; ICT = information and communication technology; n.a. = not applicable.

traceability, food safety, and sophisticated nutrition innovations is still relatively underdeveloped. Other countries in the region still need to build their STI capacity, but they have the ability to increase the adoption of some traditional and existing transformative innovations in the short term, such as less advanced or less costly e-services, urban agriculture, and precision agriculture.

AN INTEGRATED APPROACH IS NEEDED TO TRANSFORM THE REGION'S AGRI-FOOD SYSTEMS

The capacity to generate or adopt transformational innovations depends on a constellation of actors, institutions, and policies in the economy, sometimes referred to as an agricultural innovation system (AIS). A country's AIS comprises public and private agricultural R&D, extension and education organizations, farms and firms, coordination and governance arrangements, and the wider enabling environment (including regulations, infrastructure, finance) that

FIGURE ES.7

Preparedness to innovate: Schematic summary of East Asian countries' capacity to adopt technologies and to invent

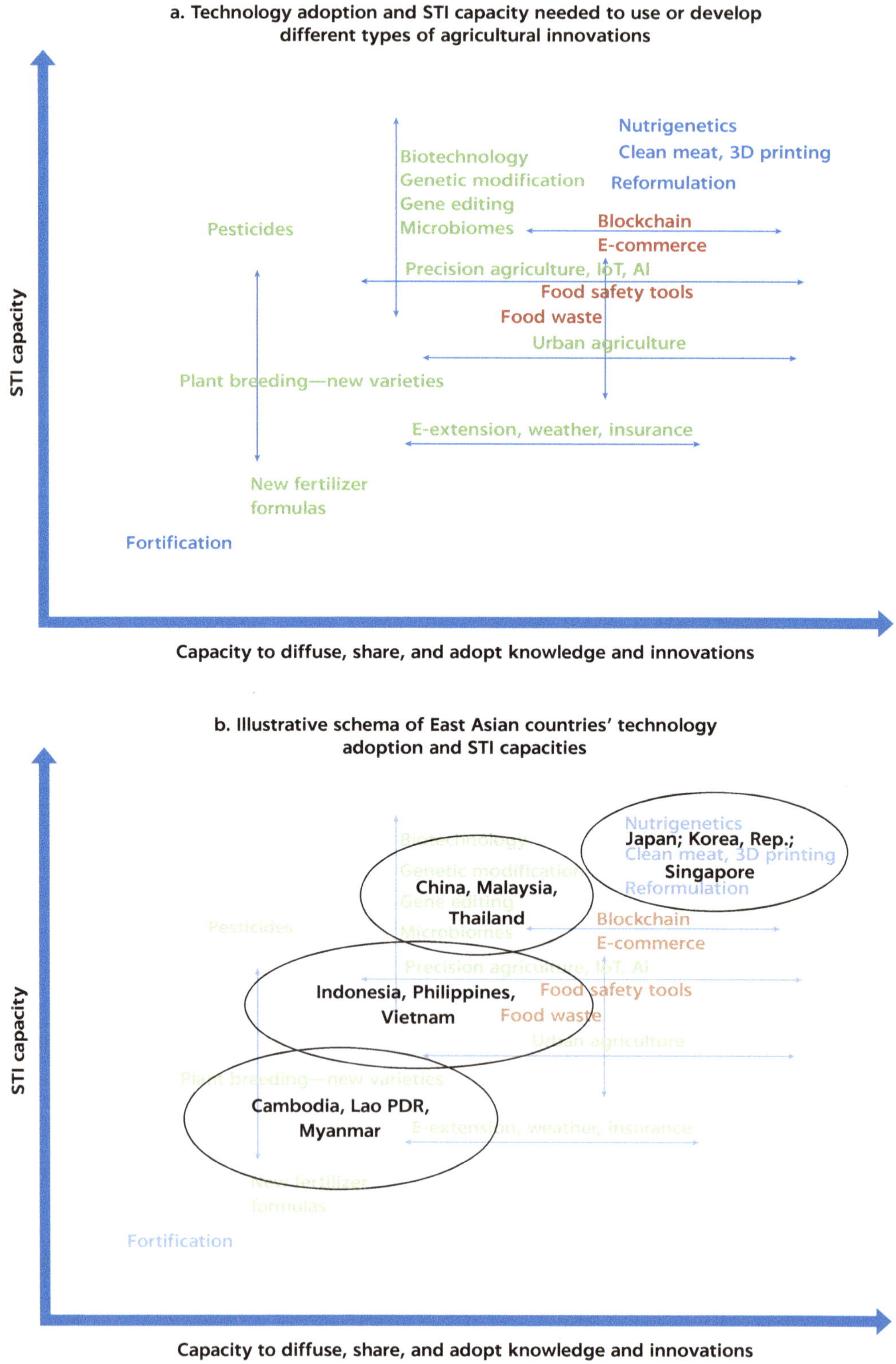

Source: Original figure for this publication, based on review of countries' innovation capabilities in agriculture.
Note: Green, red, and blue colored text refers to production, food chain, and food consumption or nutrition-oriented innovations, respectively. The arrows indicate an illustrative range of STI and diffusion and adoption capacity required to undertake the respective technology or innovation. AI = artificial intelligence; IoT = Internet of Things; STI = science, technology, and innovation.

affect innovation. The strength of a country's AIS can enable or impede innovation among farmers, firms, and public actors working within a country's agri-food system, and is thus important to determining the country's ability to use innovation as a tool for raising productivity, improving food safety, and increasing environmental sustainability.

An "AIS approach" can be instrumental to building countries' STI capacity and farmers' and firms' ability to adopt and implement transformative innovations. Policy makers in the region have not traditionally taken a holistic view of the issues affecting innovation in agri-food systems. An AIS approach can support existing efforts to transform agri-food systems by facilitating more integrated arrangements, which can help policy makers identify key constraints to innovation across the agri-food system as well as the policies and resources needed to alleviate them. It can enable them to more effectively combine traditional interventions—support for R&D, extension, and education—with complementary interventions needed to support greater innovation. Complementary interventions may include, for example, providing professional skills training through the vocational education system to support adoption of new knowledge and technologies; strengthening links between farmers and entrepreneurs, extension services, and research institutions; providing incentives for private sector–led innovation and public-private partnerships; and creating the policy, regulatory, and institutional conditions to provide incentives for and enable innovation.

Promoting innovations in the agri-food sector requires both short- and longer-term measures to strengthen each country's AIS. In the short term, reforming policies and institutional arrangements to strengthen the enabling environment for agri-food sector innovation and to guide longer-term investments in innovation capacity will be important. Where countries face resource or capacity constraints, short-term measures can be applied selectively, focusing on strengthening innovation capacity in specific segments of the agri-food sector or in specific value chains. Selective measures could include, for example, provision of training to firms and farms on produce quality and food safety standards, facilitation of public-private partnerships for applied R&D and e-extension, and targeted infrastructure investments (for example, cold chains, rural roads). Longer-term measures to build an AIS to enable transformative innovations will require a focus on strengthening human capital across the agri-food system. The short- and long-term strategies can be enacted in parallel and should complement each other. For instance, targeted short-term measures focused on specific commodities should be implemented at the same time as longer-term and more comprehensive efforts to reform agricultural training and education and demand-driven public-private R&D.

This report outlines five complementary areas of AIS investment that will be critical to spurring innovation and transforming countries' agri-food systems:

1. Strengthening innovation policy and governance
2. Reorienting agricultural R&D to meet emerging challenges
3. Modernizing agricultural extension services
4. Building skills for innovation
5. Strengthening the enabling environment for agricultural innovation

These five areas are discussed in turn.

Strengthening innovation policy and governance

Fostering transformative innovations in the region's agri-food sectors will require countries to strengthen their innovation policies. Government STI policies have traditionally focused on productivity and given cursory attention to wider agri-food system demands and challenges. In the face of rising concerns about food safety and agricultural sustainability, governments in the region will need to develop new, more modern agricultural STI policies and strategies that extend across the entire agri-food system. Such strategies should support the transition to more sustainable and resilient agriculture, which responds to growing consumer demand for safer, more nutritious food, and facilitates ongoing dietary diversification away from cereals consumption. Many countries in developing East Asia, including China, Indonesia, Malaysia, Thailand, and Vietnam, have already adopted elements of policy frameworks supporting sustainability, whereas the focus on food safety and on overall implementation of measures to enhance sustainability has been less consistent.

Fostering transformative innovations will also require strengthening AIS governance. At present, governance arrangements related to agricultural innovation in all countries in the region revolve around a dedicated organization (for example, the line ministry for the agricultural sector or a specialized institution or agency under it) with oversight responsibility for all AIS-related matters. The level of integration between the AIS and each country's broader national innovation system (NIS) varies across the region, however. In Indonesia, Malaysia, the Philippines, Thailand, and Vietnam, agriculture is integrated into NIS frameworks, that is, other ministries and agencies in the STI policy space also have a role in the governance of the AIS; in other countries, it is less integrated. Evidence reviewed for this report suggests that institutions that focus primarily on research, and not on wider innovation, are often less effective in influencing wider AIS and NIS STI policies than those with an AIS-wide mandate. Coordination across agricultural subsectors is still low, and evaluation of STI policies as part of the governance decision-making process is limited.

Countries that have moved toward nationally mandated, but independently governed, AIS organizations (as in the Republic of Korea and in Japan) have agricultural innovation policies that are better coordinated and that are more integrated with the broader NIS, which helps ensure that a more coherent set of STI policies, strategies, and activities is defined and funded. Effective processes are generally informed by stakeholder consultations and underpinned by sufficiently resourced program monitoring and evaluation efforts, which is still largely missing in all countries reviewed. The review undertaken for this report indicates that implementation and institutional ownership of AIS in the region can be strengthened through (1) stronger coordination of agricultural R&D and AIS at the national level (as in China, Indonesia, and Malaysia) and (2) greater coordination at subnational and subsectoral levels (for example, through commodity and thematic platforms in Indonesia, Malaysia, and Thailand, and via industry associations in China), using processes that include key stakeholders.

Reorienting agricultural R&D to meet emerging challenges

Aligning R&D efforts with new and emerging challenges facing the region's agri-food sector will be important. Agricultural R&D is critical both to generating innovation and to adapting technologies for adoption and implementation in local contexts. To date, agricultural R&D in the region has focused predominantly on raising cereal productivity, with relatively few resources going to the livestock or fruit and vegetable subsectors (just 3–17 percent)—products with growing demand as the region's increasingly affluent consumers diversify their diets. Shifts in consumer preferences, environmental pressures, persistent food safety and nutrition concerns, and the rising risk of zoonotic diseases all indicate that countries' agricultural R&D efforts urgently require adjustment. Although the focus on sustainability has been growing, relatively little attention has been paid to issues related to food safety, nutrition, and climate adaptation and mitigation.

Increased public and private financing of agricultural R&D, together with partnerships between public and private entities, will also be key to supporting the region's efforts to respond to emerging needs. Currently, public policy statements are incongruent with public spending allocations for AIS, including in agricultural R&D and related mechanisms for transferring knowledge to farmers (that is, extension and advisory services). Based on international experience, public and private funding for agricultural R&D is expected to be at least 1.0 percent of agricultural GDP (as in Malaysia) for countries to make sustainable progress. Investment in the range of 2.0 percent may actually be warranted if the countries aspire to join the global leaders in agri-food innovation, with a predominant share of that investment coming from the private sector. At 0.6 percent, average spending on agricultural R&D as a share of agricultural GDP in developing East Asia falls short of these benchmarks and is well below the level found in Organisation for Economic Co-operation and Development countries (2.5 percent). Private sector R&D is on the rise, representing 25 percent of all agricultural R&D spending in the East Asian upper-middle-income countries, but is still below the 50 percent share found in Organisation for Economic Co-operation and Development countries.

Private sector R&D is also an important untapped resource for agri-food sector innovation in developing East Asia. Private sector R&D tends to be geared toward modern input technologies (for example, fertilizers and agrochemicals), precision agriculture, machinery, e-services, and food-related innovations with commercial potential. Increased private sector R&D and innovation-related spending will require the removal of institutional and regulatory barriers (for example, weak enforcement of intellectual property rights [IPR], uneven playing field for domestic and foreign public and private R&D on plant breeding and GM crop development, restrictive regulations on agrochemicals, seed, and biosafety) along with the establishment of appropriate incentives for private sector participation. Although China, Malaysia, and Thailand have the institutional capacity to manage and provide legal support to IPR cases, enforcement of IPR is not yet consistent, serving as an impediment to private R&D. Supporting private sector R&D by implementing appropriate policy and regulatory reform can also create opportunities to foster co-innovation through development of public-private R&D partnerships (as in China and Indonesia). Indonesia, Malaysia, Thailand, and Vietnam have offered R&D tax credits to encourage private R&D, for example; and China has set up special economic zones to attract foreign direct investment (FDI), a large part of which goes toward R&D. Fostering the emergence of an innovation entrepreneurial community in agri-food systems will be critical if countries are to evolve from technology adopters to generators of cutting-edge innovations.

International collaboration also remains critical to addressing regionwide agri-food system challenges. The region has a long history of collaboration with the International Agriculture Research Centers, and such collaboration remains important. Evidence from global R&D reviews indicates that countries that have strengthened and leveraged domestic, regional, and international networks for knowledge exchange and collaboration on such strategic cross-country issues as climate change, natural resource management, and zoonoses are better positioned to respond to emerging needs. In short, countries in the region, even smaller ones with lower innovation capabilities, have multiple channels through which to benefit from agri-food sector R&D—building their domestic R&D capacity, enabling greater private R&D, and strengthening long-standing links to international research networks.

Modernizing agricultural extension services

To increase the impact of agri-food sector innovation on productivity, safety, and sustainability, strengthening and modernizing countries' agricultural extension services will be critical. Extension services in developing East Asian countries are still mostly public sector dominated, with variable but increasing diversity of providers, levels of decentralization, and integration of e-extension approaches. Over time, several countries in the region have increased decentralization of their extension services to the village level to improve coverage (for example, in China, Indonesia, the Philippines, and Vietnam), worked to improve the relevance and quality through more demand-driven offerings, and increased the availability and integration of public and private e-extension services (for example, rural e-centers in China and e-training in Thailand). However, concerns remain in many countries about limited coverage and low quality of both public and private extension services, inadequate penetration of e-extension (particularly in several lower-middle-income countries), chronic underfunding, and continued reliance on development partners for funding. Most extension systems still tend to be supply-driven, with weak links between knowledge providers and the farms and firms that need it, resulting in levels of technology adoption and innovation among farms and firms that are well below potential.

Agricultural extension services have the potential to increase returns to innovation, but most require reform, better integration of e-extension, and adequate and sustained funding. Evidence from global extension reviews indicates that demand-driven, pluralistic, and decentralized extension systems—ones with adequate coverage and quality, and that integrate e-extension—can better serve the needs of a diverse and often scattered clientele. Efforts to promote such extension systems may require reform of core extension policies, capacity building among both public and private service providers, and financial incentives for private and third-party service providers. Efforts to make extension services more demand-driven include supporting collective action among smallholders (as in China, Indonesia, and Thailand), establishing platforms and brokers (technology incubators and centers for technology transfer at the local and provincial levels in Vietnam), and strengthening research-extension-farm and -firm links (for example, technology demonstrations and cooperatives and industry associations in China and Indonesia). An important focus of these efforts is on enabling more systematic communication of farmer and firm demand for, and feedback on, innovations. Enabling greater integration of e-extension will require investment in improved ICT connectivity and farmer capacity, price incentives for ICT suppliers in some contexts, and training programs to upgrade the ICT skills of public and private extension agents. Delivery of extension messages can also be more effective when combined with collective action among smallholders, supporting more professional management and promoting the types of scale economies needed for the successful adoption of some transformative innovations. Finally, realizing the potential of improved extension services will require adequate and sustained funding by the region's governments.

Building skills for innovation

A lack of adequate skills of those working in the agri-food system is one of the key bottlenecks to the effective adoption of agricultural innovations in many developing East Asian countries. Although the quality of tertiary and technical and vocational education and training (TTVET) systems varies significantly across the region, they share several common characteristics. Agriculture TTVET systems in the region suffer from declining numbers of graduates because they face difficulties in attracting youth, who do not see agriculture as an attractive source of income. This challenge is accentuated by structural constraints to creating the scale required for professionally managed farming operations. Moreover, most vocational training systems are not set up to teach the skills that are needed in modern—and rapidly changing—agri-food systems. There is also limited cross-disciplinary collaboration and skills development, which constrains the capacity for co-innovation on several important (and emerging) cross-sectoral issues, including zoonoses, nutrition, and food safety. Addressing zoonoses, for example, requires collaboration across several disciplines, including agriculture and livestock, veterinary and environmental sciences, and human health. The deficiencies in the training of workers by firms and in the education and skills of an aging farm workforce limit knowledge upgrading and undermine progress toward higher-value-added production, higher productivity, and greater environmental sustainability. Several reforms in upper-middle-income and high-income countries in the region have addressed skills shortages, including by strengthening TTVET systems and fostering multidisciplinary collaboration. China, for example, has gradually reformed its agricultural tertiary education to improve both its relevance and attractiveness to farmers and firms. Korea and Singapore have improved access to and the quality of their agricultural vocational education systems to better respond to labor market needs and to prepare for the future of the sector.

Innovation capacity and skills are an essential precondition for the long-term sustainability of agri-food systems in developing East Asia. The findings of the review indicate that continued efforts to create a skilled workforce for the future across the agri-food system will require significant reform of the TTVET systems in many developing East Asian countries. The slow pace at which skills are built highlights the need for immediate action. The necessary reforms can be supported by development of a more demand-driven curriculum (content, pedagogy, cross-disciplinary courses, inclusion of industry views, industry internships) that better matches the needs of rapidly changing agri-food systems (for example, China, Malaysia, Thailand). Experience from high-income countries (for example, Australia, the Netherlands) shows that establishment of a permanent multistakeholder consultative structure that includes agri-industry, the education sector, and farm and agri-industry sector employees, and that periodically reviews the coherence and relevance of the curriculum with respect to agri-food sector needs, can be effective.

Developing East Asia needs skilled workers capable of cross-disciplinary collaboration to better respond to emerging challenges. The evidence shows that the region's TTVET systems would benefit from enhanced collaboration across disciplines and between industry and public agencies. Instruments to support improved collaboration may include programs that promote university-industry links (as in Australia, China, Malaysia, and Thailand), that establish centers of excellence (as in China, Malaysia, the Netherlands, and Thailand), that facilitate joint problem-solving (for example, in new or underfunded areas such as food safety, livestock production, and nutrition) and financing, and that increase the quality and relevance of the TTVET system. In short, the region's workforce in agribusiness and on farms is in dire need of improved knowledge and skills. Upgrading the skills of agro-industry employees is needed to improve productivity, safety, and sustainability. Although farmer capacity can be improved through extension and farmer training programs, relevant, demand-driven TTVET training becomes increasingly important when countries undergo industry-led agri-food system transformation.

Strengthening the enabling environment for agricultural innovation

The enabling environment for agricultural innovation across developing East Asia needs to be strengthened. A favorable enabling environment, especially with respect to IPR protections, food safety regulations, and infrastructure development, is critical to supporting innovation in agriculture. Adequate IPR protection is particularly important with respect to GM crops, enabling private R&D and investment, including via FDI. Most countries in the region have IPR regulations and laws in place, but the application of such regulations, for example, on plant variety protections, differs. Inadequate IPR protection has been found to impede R&D investment in GM crops (for example, in China and Indonesia). In addition to IPR, well-defined and consistently applied biosafety regulations can provide a powerful stimulus for investment in biotechnology. Most countries in developing East Asia have developed or are in the process of developing regulatory systems to manage biosafety and enable innovation in GM crops. Enforcement of biosafety regulations still faces difficulties, however, and there is limited application of GM crops in several countries (for example, Thailand and to some extent Cambodia, Lao PDR, Malaysia, and Myanmar). The study indicates that most countries in the region would benefit from aligning their IPR protection laws more closely with international standards along with strengthening their IPR enforcement. Moreover, to maximize the potential of biotechnology, establishing and improving biosafety frameworks and strengthening the enforcement of existing biosafety regulations may be warranted.

Emerging issues facing the region's agri-food systems, for example, related to food safety, are creating new challenges for policy makers in creating an adequate enabling environment. Food safety has emerged as a critical topic that is high on the agenda of consumers, governments, and the agri-food industry alike. All developing East Asian countries have food safety regulations in place; however, their capacity to respond to new issues and to enforce the regulations varies significantly. Although the more urbanized upper-middle-income countries (China, Malaysia, and Thailand) have modernized their food safety systems, most other countries covered in this study encounter significant difficulties in meeting the food safety requirements of their fast-changing food systems. The continuing gap between food safety capacity and requirements is especially challenging in rapidly urbanizing middle-income settings (as in Indonesia, the Philippines, and Vietnam). Indeed, strengthening food safety regulations and capacity, beginning with the most critical needs, such as risk-based surveillance systems, is an essential condition for adoption and invention of many food safety innovations.

Efficient and well-developed infrastructure and access to finance also play important roles in connecting farms and firms to market opportunities and enhancing the financial viability of generating and disseminating innovations. Whereas the upper-middle-income countries in the region (China, Malaysia, and Thailand) generally have good-to-excellent logistics and ICT infrastructure, lack of adequate productive infrastructure in several lower-middle-income countries (Cambodia, Indonesia, Lao PDR, Myanmar, the Philippines, and Vietnam) still limits farm and firm potential. In most developing East Asian countries, inadequate financial services available to farmers and other private actors are an investment-limiting factor. Moreover, inadequate incentives for innovation-friendly FDI still curtail the extent to which many countries benefit from innovation-linked foreign investment (Indonesia, the Philippines, and Vietnam). As such, instruments for innovation finance (for example, R&D tax credits, competitive innovation funds, and venture capital) and other support for innovation-related investments (for example, incubators, support to matchmaking, and IPR management) should be geared toward greater private sector engagement. With respect to finance, much of the region needs to improve the regulatory and institutional frameworks that govern agricultural land market operations to encourage long-term investments in agricultural productivity and sustainable growth.

A TIME FOR ACTION—TOWARD PRODUCTIVE, SAFE, AND SUSTAINABLE AGRI-FOOD SYSTEMS IN DEVELOPING EAST ASIA

The agri-food system in developing East Asia has reached a critical point. The system has provided food and employment to growing populations through innovation and economic policies that have been conducive to agricultural productivity growth. However, the region can no longer afford the economic and human costs of environmental degradation, unsafe food, persistent nutrition deficiencies, and zoonotic diseases. Surmounting these unprecedented challenges will require concerted action. Further investments in agricultural STI and in the capacity of farmers to adopt new knowledge and emerging technologies—centered on the five key areas discussed above—are urgently needed in a region that relies on millions of smallholders for their agri-food outputs and services. There is no one-size-fits-all policy mix, but countries with strong AIS capacity, national strategic leadership, and adequate financing can foster the necessary transition to more resilient, sustainable, and safe agri-food systems.

REFERENCES

Ekanayake, I. J., M. Ambrozio-Albala, and S. M. Jaffee. 2019. "From Rice to RICE: Integrated Strategies for Myanmar to Realize a More Resilient, Inclusive, Competitive and Environmentally Sensitive Food System." Presentation summarizing the "Benchmarking of Myanmar's Food Systems" report, World Bank, Washington, DC.

Scherr, Sara J., Kedar Mankad, Steven Jaffee, and Christine Negra, with Tanja Havemann, Janjarang Kijtikhun, Uray Endang Kusumjaya, Samiksha Nair, and Naomi Rosenthal. 2015. *Steps toward Green: Policy Responses to the Environmental Footprint of Commodity Agriculture in East and Southeast Asia.* Washington, DC: EcoAgriculture Partners and the World Bank.

World Bank. 2019. "Managing Headwinds." *East Asia and Pacific Economic Update* (April), World Bank, Washington, DC. doi:10.1596/978-1-4648-1412-9.

Abbreviations

3D	three-dimensional
AI	artificial intelligence
AIS	agricultural innovation system
AMR	antimicrobial resistance
ARISA	Applied Research on Innovation Systems in Agriculture
ASEAN	Association of Southeast Asian Nations
AVRDC	World Vegetable Center
AWD	alternative wetting and drying
Bt	*Bacillus thuringiensis*
CGIAR	Consultative Group on International Agricultural Research
CIAT	International Center for Tropical Agriculture
CIFOR	Center for International Forestry Research
CIMMYT	International Maize and Wheat Improvement Center
CIP	International Potato Center
CoE	center of excellence
CSA	climate-smart agriculture
DLT	distributed ledger technology
EA	East Asia
EAP	East Asia and Pacific
EID	emerging infectious disease
EU	European Union
FAO	Food and Agriculture Organization (United Nations)
FBD	foodborne disease
FDI	foreign direct investment
FFS	farmer field school
FLW	food loss and waste
GDP	gross domestic product
GE	genome/gene editing
GHG	greenhouse gas
GIS	geographic information systems
GM	genetically modified

GMO	genetically modified organism
GNI	gross national income
HEI	higher education institution
HIC	high-income country
HR	human resource
IAARD	Indonesian Agency for Agricultural Research and Development
ICRAF	World Agroforestry (formerly International Council for Research in Agroforestry)
ICRISAT	International Crops Research Institute for Semi-Arid Tropics
ICT	information and communication technology
ILRI	International Livestock Research Institute
IoT	Internet of Things
IPM	integrated pest management
IPR	intellectual property rights
IRRI	International Rice Research Institute
IWMI	International Water Management Institute
LIC	low-income country
MG	matching grant
MIC	middle-income country
MV	modern variety
NBT	new breeding techniques
NIS	national innovation system
PCARRD	Philippine Council for Agriculture, Aquatic, and Natural Resources Research and Development
PPP	public-private partnership
PTD	participatory technology development
RFID	radio-frequency identification technology
SEZ	special economic zone
SME	small and medium enterprise
SOE	state-owned enterprise
SRI	sustainable rice intensification
STI	science, technology, and innovation
TFP	total factor productivity
TRIPS	Trade-Related Aspects of Intellectual Property Rights
TTO	technology transfer office
TTVET	tertiary technical and vocational education and training
UPOV	International Convention for the Protection of New Varieties of Plants
VAAS	Vietnam Academy of Agricultural Sciences
WIPO	World Intellectual Property Organization
WUA	water user association
WUG	water user group

All dollar amounts are US dollars unless otherwise indicated.

1 Status of Agri-Food Systems in East Asia

THE AGRI-FOOD SECTOR AND OVERALL DEVELOPMENT IN DEVELOPING EAST ASIA

The agri-food sector is of major importance for the developing East Asian countries, providing nutrition and livelihoods and generating both employment and economic output. The food value chain contributes to roughly 14 percent of the region's gross domestic product (GDP) and 35 percent of the labor force, of which about 60 percent are smallholders. The share of agriculture in overall national employment and GDP has, however, been declining as broad-based economic development has progressed (table 1.1). Yet agriculture-related employment is expected to remain prominent in most countries of the region.

Rising agricultural productivity has contributed to impressive reductions in poverty, and the countries have mostly met their food security needs[1] (table 1.1). Although moderate food insecurity is still widely prevalent in some countries in the region, severe food insecurity has been significantly reduced. The achievements in productivity are largely attributable to progress in crop science, favorable government policies, and the delivery of extension services to smallholders (Fuglie et al. 2020; World Bank 2019). Agricultural performance remains important in achieving further poverty reduction gains (Briones and Felipe 2013; Timmer 2014), particularly among the lower-middle-income countries (lower MICs) of Cambodia, the Lao People's Democratic Republic, and Myanmar.[2] Some of the region's food demand cannot be competitively supplied locally, resulting in increased food, feed, and live animal imports (ACI 2010). Most of the region's MICs have become net importers of crop and livestock products (table 1.2).

The relative value of the food sector has increased in many countries (figure 1.1), whereas agricultural GDP relative to total GDP has decreased over the years (table 1.1). The value lies increasingly in post–farm gate activities (for example, logistics, processing, packaging, and retail). Prices to consumers are highly influenced by conditions in the supply chain beyond the farm gate (for some 50–70 percent of the cost of food)[3] (Reardon and Timmer 2014).

TABLE 1.1 **Agriculture sector and development statistics**

	AGRICULTURE VALUE ADDED (% OF GDP)			AGRICULTURE EMPLOYMENT (% OF WORKFORCE)		AGRICULTURE GDP (ANNUAL GROWTH %)[a]			POVERTY (% OF POPULATION)		FOOD INSECURITY, 2016–18 (% OF POPULATION)[b]	
COUNTRY	1961	1991	2018	1991	2018	1961–80	1981–2000	2001–18	1990[c]	2015–16	MODERATE TO SEVERE	SEVERE
Cambodia	—	45.3 (1993)	22.0	75.7	30.0	—	4.27	3.65	—	22.7	44.9	14.2
China	35.8	24.0	7.2	59.7	26.8	3.57	5.0	3.98	66.2	0.5	9.6	1.1
Indonesia	24.1 (1983)	19.7	12.8	54.0	30.5	3.10	2.84	3.70	66.7 (1998)	6.5	8.1	1.0
Lao PDR	—	44.0	15.7	86.1	67.7	—	—	3.14	52.4 (1997)	2.0	—	—
Malaysia	45.4	14.4	7.5	22.0	11.1	4.80	2.09	2.60	1.6 (1989)	0	—	—
Myanmar	—	57.2 (2000)	24.6	69.5	49.7	—	—	4.74	—	—	—	—
Philippines	26.8	21.0	9.3	45.2	25.2	4.19	1.64	2.26	13.9 (2000)	6.1	52.5	15.0
Thailand	35.8	12.7	8.1	60.3	30.7	4.80	3.11	1.80	9.4	0	—	—
Vietnam	—	40.5	14.7	68.6	39.4	—	3.74	3.23	52.9 (1992)	2.0	14.5	2.3

Sources: Original table for this publication. FAOSTAT (Statistics Division, Food and Agriculture Organization of the United Nations), 2020 (http://www.fao.org/faostat/en/#data); World Bank Databank, 2020 (https://databank.worldbank.org/home.aspx).
Note: — = not available.
a. Data are not available for all years in a given period; some values presented are averages for available yearly data.
b. Food insecurity (2016–18) is prevalence of moderate to severe food insecurity in the total population.
c. Values for different years: Malaysia (1989), Indonesia (1998), Philippines (2000), Lao PDR (1997), Mongolia (1995), the rest (1990).

FIGURE 1.1
The relative value of the food sector in selected East Asian countries, 2015

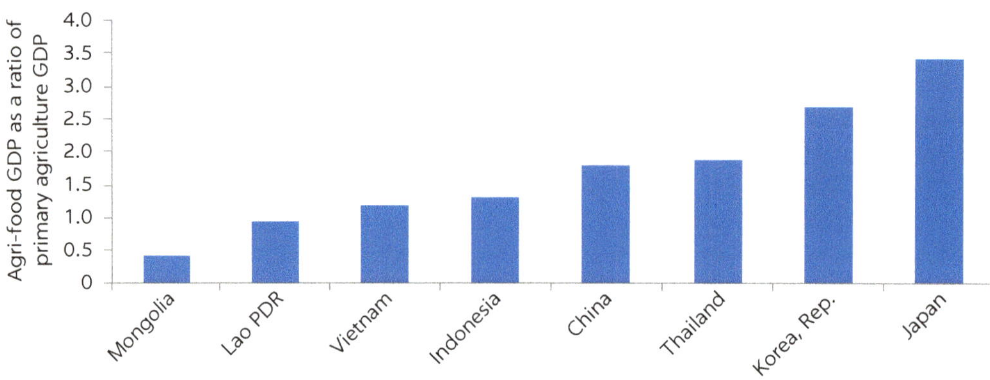

Source: World Bank 2017.

Agricultural commodity exports have exploded during the past 20–30 years (table 1.2), causing East and Southeast Asia to dominate world trade in rubber, palm oil, black pepper, and cassava.[4] Exports of high-value food safety–sensitive products, such as fish, poultry, and horticulture, grew from $22.6 billion in 2001 to $97.3 billion in 2017 (World Bank 2019). Although changes in Chinese agriculture have primarily been driven by changing domestic food market demand, much of the dynamics in the agricultural sectors of other upper and mid-range MICs—Indonesia, Malaysia, Thailand, and Vietnam—have been related to

TABLE 1.2 **Agricultural sector economic performance**

	AGRICULTURAL TFP (AVERAGE GROWTH %)		AGRICULTURE VALUE ADDED PER WORKER (CONSTANT 2010 $)		AGRICULTURAL EXPORTS ($, MILLION)		AGRICULTURAL IMPORTS ($, MILLION)		FDI TO AGRICULTURE (% OF TOTAL FDI)[a]	FDI TO FOOD AND BEVERAGE (% OF TOTAL)
COUNTRY	1991–2000	2006–10	1991	2018	1961	2017	1961	2017	2010, 2014	2010
Cambodia	—	—	552 (1994)	1,597	60.8	996	13.1	2,664	11.5	0.35
China	4.13	3.25	714	3,830	267	53,330	666	117,411	2.33 (2014) 1.48 (2010)	1.68
Indonesia	1.23	2.62	—	1,379 (2015)	465	39,696	146	19,692	2.45	—
Lao PDR	—	—	585 (2000)	785	0.2	696	11.7	1,115	—	—
Malaysia	1.87	2.94	11,534	18,008	615	23,778	312	16,587	–0.16	—
Myanmar	—	—	665 (2000)	1,713	187.6	2,332	32.7	3,999	10.0	—
Philippines	0.46	1.68	1,496	2,557	364	5,674	110.2	10,710	0.18	—
Thailand	3.27	1.60	1,412	3,344	392	32,169	46.8	11,372	0.0	1.53
Vietnam	2.86	2.18	450	1,212	72.7	18,774	40.8	22,506	1.0 (2014) 0.45 (2010)	—

Sources: Original table for this publication. FAOSTAT (Statistics Division, Food and Agriculture Organization of the United Nations), 2020 (http://www.fao.org/faostat/en/#data); World Bank 2016; World Bank Databank, 2020 (https://databank.worldbank.org/home.aspx).
Note: — = not available; FDI = foreign direct investment; TFP = total factor productivity.
a. China, Malaysia, and Thailand also had FDI outflows ($534.0 million, $433.0 million, and $0.3 million, respectively). China's net inflow of FDI to agriculture (2010) was $1,700.7 million, whereas for most other countries it ranged from a few million to $156.0 million and $337.0 million for Cambodia and Indonesia, respectively. Myanmar receives some FDI to agriculture, and it will likely receive greater net inflows of FDI in the future (palm oil and rubber). FDI values before 2010 are not available.

commodity production for external demand (Leimona et al. 2015; Nguyen et al. 2015; Wicke et al. 2011). Lower MICs, such as Cambodia and Myanmar, have also emphasized the potential of commodities trade in their evolving agricultural strategies (Eliste and Jaffee 2014).

Over the years, developing East Asia's export trade has changed in both composition and target markets. The composition has shifted away from traditional tropical products, such as coffee, cocoa, tea, sugar, spices, and nuts, toward products such as horticulture, seafood, and processed products. Since the early 2000s, countries have endorsed increasingly different export strategies targeting either processed (Indonesia, Malaysia) or nonprocessed (Thailand, Vietnam) food exports, depending on their comparative advantage (UN ESCAP 2011). Since the early 2000s, a growing proportion of this trade has been redirected to other emerging market countries, including within Asia (Scherr et al. 2015). East and Southeast Asian countries are also working to deepen regional economic integration, partly to offset increasing global protectionism and weakened demand for East Asia's exports (IFPRI 2020).

DRIVERS OF EAST ASIA'S AGRI-FOOD SYSTEM TRANSFORMATION: CHANGING SOCIOECONOMIC NEEDS AND EMERGING EXTERNAL RISKS

Agri-food systems in developing East Asia are undergoing rapid structural transformation, largely driven by economic growth patterns, rapid urbanization with changing diets and demand for more nutritious and safe food, and accompanying pressures on natural resources (Scherr et al. 2015; World Bank 2019). Domestic dietary and food expenditure patterns are changing because of higher incomes and greater formal labor participation, reinforced by accelerating urbanization. The prominence in diets of staples such as rice is being reduced, whereas the importance of other cereals and higher-value fruits, vegetables, fish, animal products, and packaged convenience foods has increased rapidly (Jamora 2014). The majority of consumers are expected to become fully reliant on purchased foods and ingredients.[5] These demand shifts have also altered specialization patterns among smallholder farmers and larger food producers and have spurred changes in agricultural land and water use. The region's diversifying food needs and growing urban population[6] are now being supplied by a hybrid system that combines integrated, advanced cold chain distribution channels with marketing networks that still rely on traditional standards and distribution outlets to supply sophisticated and vertically integrated supermarket food chains driven by domestic and international firms.

Land availability is one of the key constraints limiting the ability of agri-food systems to respond to food demand. Continued conversion of the most productive agricultural land to other uses, such as urban areas (table 1.3), industry, and biofuels (sourcing, for example, cereals, oilseeds, and oil palm) has exacerbated the productivity pressures on food systems (d'Amour et al. 2017; Scherr et al. 2015). The projected yield declines caused by land degradation linked to water shortages, poor soil management, and climate change compound the challenges from competing land use. Land availability is expected to worsen as the effects of climate change are increasingly felt. Because of desertification and salinization, the amount of arable land per capita may fall by as much as 5 percent by 2030 (chapter 2) (PwC, Rabobank, and Temasek 2019).

The low availability of land is compounded by stagnating yields, small and fragmented farm structures, and aging farmers. Productivity growth is stalling in

TABLE 1.3 **Cropland and crop production loss in selected East Asian countries**

COUNTRY OR AREA	EXPECTED CROPLAND LOSS (MILLION HECTARES)[a]	SHARE OF TOTAL CROPLAND (%)	SHARE OF TOTAL CROP PRODUCTION (%)	LAND PRODUCTIVITY COMPARED WITH NATIONAL AVERAGE[b]
China	7.6	5.4	8.7	1.53
Indonesia	0.6	1.1	2.3	2.03
Philippines	0.3	2.9	4.8	1.66
Vietnam	0.8	10.3	15.9	1.41
Asia	18.0	3.2	5.6	1.59
World	30.0	2.0	3.7	1.77

Source: Adapted from d'Amour et al. 2017.
a. Projection based on estimated land loss to population growth and urbanization from 2000 to 2030 near large urban areas.
b. Urban expansion is expected to take place on cropland that is 1.77 times more productive than the global average. Productivity calculations use mapping and data mapping on four staple and three cash crops.

several countries. In China and Indonesia, for example, rice yield increases began to taper off in the 1990s as years of environmentally damaging practices took their toll on soil and water resources (Alexandratos and Bruinsma 2012; APEC 2014; Garnett and Wilkes 2014; Rabobank Economic Research 2016). The region has fallen behind many of the other regions of the world; for instance, despite the volumes of rice that are currently exported by many East Asian countries, the yield gap with the United States and Brazil still stands at about 40 percent and 20 percent, respectively (PwC and FIA 2020). It will be increasingly difficult to achieve high yields given the multiple threats introduced by climate change (such as pests, droughts, and flooding). The region's small and fragmented farm structure is not amenable to many productivity-enhancing investments (for example, mechanization) (chapter 5). Moreover, the region's aging farmers (in China, Malaysia, the Philippines, Thailand, and Vietnam) (OECD 2016, 2018; OECD and World Bank 2014; UNCTAD 2018) have typically attained lower levels of education, which has been found to impede, among other things, adoption of new productivity-enhancing technology (chapter 6).

The supply-side challenges are further compounded by the high level of food loss and waste in the region. Food loss and waste are driven by a combination of fragmented supply chains; underinvestment in infrastructure, such as cold storage systems; and growing consumer waste (Ecosperity 2018; PwC and FIA 2020). Globally, about 30 percent of food produced is currently wasted between the farmer and the consumer (FAO 2017), giving rise to high levels of greenhouse gas emissions.

Amid the region's socioeconomic changes, the agri-food system in developing East Asia has also become increasingly vulnerable to a number of risks and external shocks. The agri-food system has been under growing pressure to address the economic, environmental, and health-related weaknesses associated with expanding and intensifying production (chapter 2), and has been gripped with reduced land availability, falling productivity growth and stagnating yields, climatic and weather-related risks, and emerging infectious diseases of zoonotic origin. The COVID-19 (coronavirus) outbreak has further exposed the weaknesses in food systems by disrupting supply channels and disconnecting producers from markets (chapter 2). Such supply-side challenges further constrain the agri-food sector's ability to meet the growing demand for food imposed by population growth.[7]

Advancements in the agri-food sector are critical for its continued performance. The agri-food sector in the region faces unprecedented challenges in meeting the region's growing demand for food. These challenges heighten the urgency of ensuring the sector's performance, including continuing total factor productivity growth (table 1.2), and sustainable transformation of the region's agri-food systems. Further investments in agricultural science, technology, and innovation by both the public and private sectors, and in management practices, capacity, policy, and governance are direly needed in a region that relies on millions of smallholders for its agri-food outputs and services.

THE PURPOSE OF THE REPORT

This report assesses the institutional readiness of the developing East Asian countries to generate and adopt agricultural innovations that have the potential to redirect the transformation of agri-food systems toward the triple win of productivity, sustainability, and health. The report focuses primarily on China, Malaysia, and Thailand (the upper MICs); Indonesia, the Philippines, and Vietnam (the mid-range MICs); and to some extent Cambodia, Lao PDR, and Myanmar (the lower MICs). The report first reviews the status of the agri-food system in developing East Asia (chapters 1 and 2) and discusses the role of innovation and agricultural innovation systems (AIS) in transformation and growth (chapter 3) (table 1.4). AIS refers to a network of relevant public and private actors collaborating within a wider policy, governance, and enabling environment. Second, the report reviews past and present innovations and their outcomes (looking backward; chapter 4), and transformative innovations and their potential to address diverse agri-food system needs (looking forward; chapter 5). Selected countries' readiness to embrace these transformative innovations is briefly reviewed and assessed in chapter 5.

Finally, the report reviews the countries' overall AIS capacity (chapter 6). This volume is a desk review and limited by evidence and the concepts of

TABLE 1.4 **Navigating the report**

CHAPTER 1: STATUS	**CHAPTER 4: LOOKING BACKWARD**	**CHAPTER 5: LOOKING FORWARD**	**CHAPTER 6: INNOVATION CAPACITY**
Agri-food sector performance in developing East Asia Drivers for transformation Key constraints limiting supply response	**Food Security and Exports** Green Revolution technologies Plantation and aquaculture production	**Agricultural Production** Technology focus: ICT: precision agriculture, e-services, blockchain-enabled e-services Biotechnology: genetic modification, gene editing, microbiomes, feed additives	Country vision, governance, and leadership
CHAPTER 2: CHALLENGES	**Sustainable Practices**	Off-farm focus: Urban (and protected) agriculture	Agricultural research: public, private, international collaboration
Environmental degradation and economic health	Integrated cropping practices		
Food safety and nutrition	Input management	**Food Chain**	Agricultural extension services
Zoonoses management and impact of COVID-19	Sustainable and carbon-neutral livestock production	Blockchain for traceability and safety Food sensing technology	
CHAPTER 3: INNOVATION AND AIS	Genetically modified crops	Packaging, food loss and waste, e-commerce	Incentives for private sector–led innovation
Country categories—agrarian, transitioning, and urban	Organic agriculture	**Consumption and Nutrition**	Innovation skills across the agri-food system
Agricultural innovation process		Fortification and biofortification, reformulation, functional foods, nutrigenetics	
AIS framework	Irrigation and water management	Alternative food sources **Potential of Innovations**	AIS governance
Holistic AIS approach and the value added to existing efforts to transform agri-food system and zoonoses management	Climate-smart agriculture practices	Economic, environmental, health, and social potential of innovations (table 5.1) Trade-offs	Enabling environment
	Adoption and scaling up	Country readiness to embrace innovations (figure 5.3; tables 5.2 and 5.3)	**CHAPTER 7: CONCLUSIONS**

Source: Original table for this publication.
Note: AIS = agricultural innovation system; ICT = information and communication technology.

AIS used in any given report. The review draws on existing literature, including reports by the International Food Policy Research Institute, the Organisation for Economic Co-operation and Development, and the World Bank, but does not extensively repeat the findings in them. Only a few elements of AIS (table 1.4) are addressed more thoroughly, and to a large extent descriptively. The assessment does not replace due diligence on analyzing a country's AIS.

NOTES

1. The East Asia and Pacific population consuming an inadequate level of calories fell from 23 percent to 11 percent between 1990–92 and 2012–14 (World Bank 2019).
2. The report covers three groups of East Asian countries: China, Malaysia, and Thailand (the upper MICs); Indonesia, the Philippines, and Vietnam (the mid-range MICs); and to some extent Cambodia, Lao PDR, and Myanmar (the lower MICs).
3. For example, the rice farming segment constitutes only 10 percent of the total value of food in the Asian economy (Reardon and Timmer 2014).
4. Several countries in East Asia lead (for details, see chapter 4) in the world's exports of rice (China, Thailand, Vietnam), coffee (Indonesia, Lao PDR, Vietnam), tea (China), cocoa (Indonesia, Malaysia), cashews (Vietnam), aquaculture (China, Indonesia, Vietnam), fruits (Philippines, Thailand, Vietnam), and vegetables (China, Vietnam) (Scherr et al. 2015)
5. Growth in per capita consumption of processed foods (1999–2017): large increases in China (9.4 percent), Vietnam (7.8 percent), Indonesia (5.5 percent), and Thailand (4.4 percent); slight decreases in Japan (0.5 percent) and the Philippines (0.1 percent) (Jaffee 2019).
6. In 2010, the East Asia and Pacific urban population amounted to 754 million people. Estimates are that by 2030 an additional 550 million people will have moved to cities in the region (Ecosperity 2018).
7. The global population is forecast to increase to 10 billion by 2050, requiring about a 50 percent increase in food production to meet demand (FAO 2018).

REFERENCES

ACI (Agrifood Consulting International). 2010. "Scoping Study on the Challenges for the Animal Feed Complex in East Asia." Report for the World Bank, Agrifood Consulting International, Kensington, MD.

Alexandratos, N., and J. Bruinsma. 2012. "World Agriculture towards 2030/2050: The 2012 Revision." ESA Working Paper 12-03, Food and Agriculture Organization of the United Nations, Rome.

APEC (Asia-Pacific Economic Cooperation). 2014. "APEC Food Security Roadmap towards 2020 (Version 2014 Draft Track)." Paper for the Policy Partnership on Food Security Meeting, Beijing, China, August 14–15.

Briones, J., and J. Felipe. 2013. *Agriculture and Structural Transformation in Developing Countries in Asia: Review and Outlook*. Mandaluyong City: Asian Development Bank.

d'Amour, C. B., F. Reitsmac, G. Baiocchi, S. Barthelf, B. Güneralp, K.-L. Erb, H. Haberl, F. Creutzig, and K. C. Seto. 2017. "Future Urban Land Expansion and Implications for Global Croplands." *PNAS* 114 (34): 8939–44.

Ecosperity. 2018. "Feeding Urban Asia: New Approaches for Providing Safe, Nutritious, and Affordable Food." Ecosperity, Singapore.

Eliste, P., and S. Jaffee. 2014. "Re-entering the Game: Policy Options for Emerging Rice Exporters." In *Rice in the Shadow of Skyscrapers: Policy Choices in a Dynamic East and Southeast Asian Setting*, edited by D. Dawe, S. Jaffee, and N. Santos. Rome: Food and Agriculture Organization of the United Nations.

FAO (Food and Agriculture Organization of the United Nations). 2017. "Save Food: Global Initiative on Food Loss and Waste Reduction." FAO, Rome. http://www.fao.org/save-food/resources/keyfindings/en/.

FAO (Food and Agriculture Organization of the United Nations). 2018. *The Future of Food and Agriculture: Alternative Pathways to 2050. A Summary Review.* Rome: FAO.

Fuglie, K., M. Gautam, A. Goyal, and W. F. Maloney. 2020. *Harvesting Prosperity: Technology and Productivity Growth in Agriculture.* Washington, DC: World Bank.

Garnett, T., and A. Wilkes. 2014. *Appetite for Change: Social, Economic and Environmental Transformations in China's Food System.* Oxford, UK: Food Climate Research Network.

IFPRI (International Food Policy Research Institute). 2020. *Building Inclusive Food Systems. IFPRI 2020 Global Food Policy Report.* Washington, DC: IFPRI. http://ebrary.ifpri.org/utils/getfile/collection/p15738coll2/id/133646/filename/133857.pdf.

Jaffee, S. 2019. "Emerging Asia's Safe Food Imperative. Towards Smarter Investment and Regulatory Delivery." Presented at Food Industry Asia/World Bank Group Seminar, "Accelerating Food Safety in Asia," Singapore, February 19.

Jamora, N. 2014. "Overview of Food Demand Trends and Prospects in East Asia." Report prepared for the World Bank, Georg-August-University of Göttingen.

Leimona, B., S. Amaruzaman, B. Arifin, F. Yasmin, F. Hasan, B. Dradjat, H. Agusta, S. Jaffee, and J. Frias. 2015. *Indonesia's "Green Agriculture" Strategies and Policies: Closing the Gap between Aspirations and Application.* Occasional Paper 23. Nairobi: World Agroforestry Centre.

Nguyen, D. A. T., A. P. Nguyen, K. V. Chinh, and C. N. Nguyen. 2015. "Vietnam Country Study." Unpublished working paper, World Bank, Washington, DC.

OECD (Organisation for Economic Co-operation and Development). 2016. *OECD Reviews of Innovation Policy: Malaysia 2016.* Paris: OECD Publishing. http://dx.doi.org/10.1787/9789264255340-en.

OECD (Organisation for Economic Co-operation and Development). 2018. *Innovation, Agricultural Productivity and Sustainability in China.* OECD Food and Agricultural Reviews. Paris: OECD Publishing. https://doi.org/10.1787/9789264085299-en.

OECD (Organisation for Economic Co-operation and Development) and World Bank. 2014. *Science, Technology and Innovation in Viet Nam.* Paris: OECD Publishing. http://dx.doi.org/10.1787/9789264213500-en.

PwC (PricewaterhouseCoopers) and FIA (Food Industry Asia). 2020. "Maintaining Food Resilience in a Time of Uncertainty: Understanding the Importance of Food Value Chains in ASEAN and How to Ensure Their Resilience during the COVID-19 Crisis." PricewaterhouseCoopers, Commissioned by Food Industry Asia.

PwC (PricewaterhouseCoopers), Rabobank, and Temasek. 2019. *The Asia Food Challenge: Harvesting the Future.* Report prepared for 2019 Asia-Pacific Agri-Food Innovation Week, Singapore, November 20–22. https://www.asiafoodchallenge.com/documents/afc-report-112019.pdf.

Rabobank Economic Research. 2016. "Asia-Pacific: Agricultural Perspectives." Rabobank Economic Research, Utrecht. https://economics.rabobank.com/publications/2016/february/asia-pacific-agricultural-perspectives/.

Reardon, T., and P. Timmer. 2014. "Five Inter-Linked Transformations in the Asian Agrifood Economy: Food Security Implications." *Global Food Security.* 3 (2): 108–17. https://doi.org/10.1016/j.gfs.2014.02.001.

Scherr, Sara J., Kedar Mankad, Steven Jaffee, and Christine Negra, with Tanja Havemann, Janjarang Kijtikhun, Uray Endang Kusumjaya, Samiksha Nair, and Naomi Rosenthal. 2015. *Steps toward Green: Policy Responses to the Environmental Footprint of Commodity Agriculture in East and Southeast Asia.* Washington, DC: EcoAgriculture Partners and the World Bank.

Timmer, P. 2014. "Managing Structural Transformation: A Political Economy Approach." Presented at 18th Annual WIDER Lecture, United Nations Headquarters, New York, November 18.

UN ESCAP (United Nations Economic and Social Commission for Asia and the Pacific). 2011. *Facilitating Agricultural Trade in Asia and the Pacific.* ESCAP Studies in Trade and Investment 74. New York: United Nations. https://doi.org/10.18356/8da51b4c-en.

UNCTAD (United Nations Conference on Trade and Development). 2018. *Science, Technology & Innovation Policy Review. Thailand.* Geneva: UNCTAD.

Wicke, B., R. Sikkema, V. Dornburg, and A. Faaij. 2011. "Exploring Land Use Changes and the Role of Palm Oil Production in Indonesia and Malaysia." *Land Use Policy* 28: 193–206.

World Bank. 2016. *Transforming Vietnamese Agriculture: Gaining More from Less.* Vietnam Development Report 2016. Washington, DC: World Bank.

World Bank. 2017. "Megatrends and the Trajectory of Agro-food Systems in East Asia." Agriculture and Food Forum, "*Tomorrow's Agriculture–Starting Today.*" Tyson's Corner, VA, March 6–10.

World Bank. 2019. *Managing Headwinds. East Asia and Pacific Economic Update.* Washington, DC: World Bank. doi:10.1596/978-1-4648-1412-9.

2 Challenges Facing Agri-Food Systems in East Asia

THREATS TO THE AGRI-FOOD SYSTEM'S PRODUCTIVITY AND SUSTAINABILITY

The success of the agriculture sector in developing East Asia is associated with a heavy environmental footprint. Despite the sector's contributions to economic development in the region, continued emphasis on food security, productivity, and rapid commodity growth has had major negative impacts on the environment. These environmental impacts have at times negatively affected agricultural productivity,[1] health, and livelihoods. Although agricultural pollution has progressed at different rates across the region, it has become a concern in all countries where farming has taken an intensive turn. Several factors drive these impacts, such as conversion of new agricultural land, limited farmer knowledge of pesticide use[2] and good farming practices, and weak institutional support for sustainability standards. These practices and limitations have led to deforestation; soil erosion; water, air, and marine plastics pollution[3] (chapters 4 and 5); accelerated release of greenhouse gases (GHGs); water scarcity; and the salinization of water resources. The expansion or intensification of commodity agriculture may also pose threats to biodiversity, create competition for scarce water resources, and contribute to climate change (Scherr et al. 2015; World Bank 2004) and the emergence of zoonoses (McAleenan and Nicolle 2020).

The costs associated with environmental degradation and adverse impacts are substantial. At the individual farm and community levels the costs arise from, for example, decreasing productivity, lower production, and a wide variety of health effects (EJF 2003; Sapkota et al. 2008). National economic losses associated with environmental degradation are also significant across East and Southeast Asian countries (Scherr et al. 2015). For instance, natural resources depletion has been estimated to range from 2 percent of gross national income in the Philippines to 12 percent of gross national income in the Lao People's Democratic Republic (figure 2.1). In the East Asia and Pacific Region, the annual cost of plastic pollution to the tourism, fishing, and shipping industries alone has been estimated at $1.3 billion (World Bank 2019b).

FIGURE 2.1

Natural resources depletion in developing East Asia

Cost as a percent of gross national income

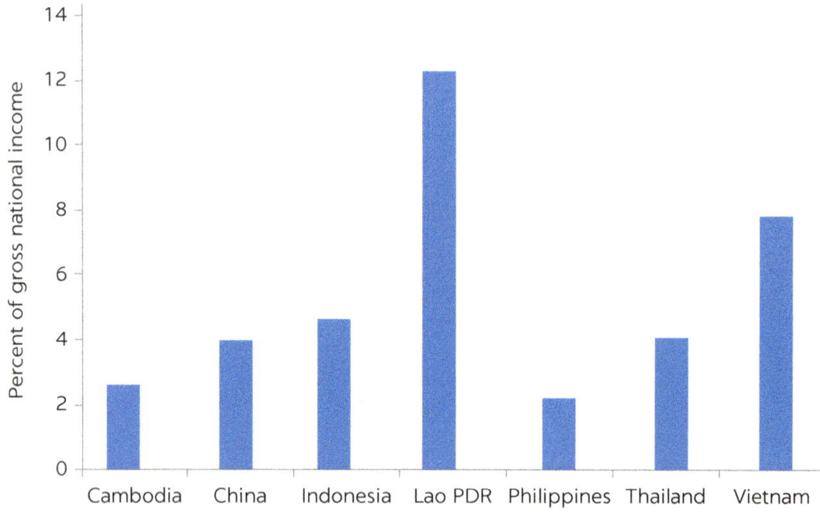

Source: Adapted from Scherr et al. 2015.

Note: Natural resources depletion (2012 data) is the sum of net forest depletion, energy depletion, and mineral depletion. Estimates exclude the economic costs of greenhouse gas damage as well as ecosystem service losses.

The nature and relative importance of environmental risks, however, vary for different commodities and farming systems. The major adverse impacts of agriculture in East Asia are summarized in table 2.1. Whereas rice and sugarcane pose a wide range of environmental impacts, the impacts from coffee production are greatest for soil degradation and water scarcity. The production of rubber carries the highest risk factor for biodiversity loss. The actual impacts depend on production practices; patterns of land use, including land conversions; and local environmental conditions, and can be mitigated or even reversed through good farm management and land-use practices (Scherr et al. 2015).

Climate change also results in significant impacts on the region's food production, and the region's countries are likely to be severely affected. Climate change has created new pressures on productivity and the resilience of the agri-food system. Climate change will also have an impact on public health outcomes. For instance, endemic zoonotic diseases, which are amplified by climate change, are posing increasing risks.[4] Table 2.2 presents the estimated effect of climate change on rice yields in selected locations and lists the sector's main GHG emission sources.[5] For example, the Food and Agriculture Organization of the United Nations estimates a 3–10 percent decline in average global cereal yields for every 1°C increase (Fuglie et al. 2020). Reduced rainfall will deplete water resources, and increasing temperatures will disturb crop development cycles, rendering some unviable. Rising sea levels and flooding are already threatening currently fertile geographies, including China's wheat- and maize-growing regions, such as the Pearl River Delta, the Yangtze River Delta, and coastal Jiangsu Province; Vietnam's rice fields in the Mekong Delta and Red River Delta; and large swathes of low-lying countries, such as Indonesia (FAO 2015).

Although other sectors, such as energy and transport, have seen their GHG emissions decline, agriculture-related GHGs are still increasing (Acosta et al. 2019). To prevent agriculture's share of global GHGs from increasing from the

TABLE 2.1 **Principal environmental risks, by commodity, in East and Southeast Asia**

COMMODITY	GEOGRAPHIC AREA	SOIL DEGRADATION	POLLUTION	WATER SCARCITY OR SALINIZATION	BIODIVERSITY AND HABITAT LOSS[a]	GHG EMISSIONS[b]
Coffee	Tropical uplands	Dark blue	Yellow	Dark blue	Light blue	Yellow
Cocoa	Tropical uplands	Light blue	Dark blue	Yellow	Light blue	Yellow
Rubber	Tropical uplands	Yellow	Light blue	Yellow	Dark blue	Yellow
Bananas	Tropical uplands	Dark blue	Dark blue	Light blue	Yellow	Yellow
Palm oil	Lowlands/midlands	Dark blue	Light blue	Yellow	Dark blue	Dark blue
Tea	Tropical uplands	Light blue	Dark blue	Yellow	Dark blue	Yellow
Maize	Varied	Dark blue	Dark blue	Dark blue	Light blue	Light blue
Rice	Varied	Dark blue	Dark blue	Dark blue	Dark blue	Dark blue
Sugarcane	Lowlands	Dark blue	Dark blue	Dark blue	Yellow	Dark blue
Pork (intensive)	Varied	Yellow	Dark blue	Yellow	Yellow	Dark blue
Shrimp (intensive)	Coastal lowlands	Yellow	Dark blue	Yellow	Dark blue	Yellow

Source: Scherr et al. 2015.
Note: Dark blue signifies high risk, light blue signifies medium risk, and yellow signifies low risk.
a. Including forest clearing.
b. Greenhouse gas (GHG) emissions from land uses and land clearing.

TABLE 2.2 **Expected yield decrease and agriculture-related greenhouse gas emissions in East Asia**

COUNTRY OR AREA	EXPECTED YIELD DECREASE[a] IN RICE (%)	AGRICULTURE-RELATED GHG EMISSIONS (% OF TOTAL)[b]		
		LIVESTOCK	RICE	FERTILIZER
China	4.0	29.0	50.0	36.0
Myanmar	5.0	57.0	45.5	1.5
Philippines	10.0	25.0	63.0	8.0
Vietnam	4.3	32.0	68.0	14.0
Asia	—	4.3	5.3	—
World	3.0–10.0	—	—	—

Sources: World Bank. Estimates from Climate Smart Agriculture Profiles for Vietnam and the Philippines; Ekanayake, Ambrosio-Albala, and Jaffee 2019 (Myanmar); Fuglie et al. 2020; Searchinger 2016 (China); World Bank 2017 (Myanmar).
Note: — = not available or appropriate; GHG = greenhouse gas.
a. Estimated yield decreases by 2045–50 for each 1°C increase in temperature.
b. Fertilizer value includes emissions from fertilizer use for all crops. Rice emissions also include emissions from rice-specific fertilizer use.

current 49 gigatons (Gt), while reducing the overall agriculture-related GHG emissions to 21–22 Gt, would require a 60 percent reduction of agriculture-generated GHGs by 2050 (WRI 2013). In the East Asia and Pacific Region, the main sources of agriculture-related GHGs are overuse of nitrogen fertilizer and methane emissions from livestock and rice production systems. Most countries have endorsed agricultural production programs that address mitigation (for example, through measures that reduce nitrogen overuse and improve water management in rice and livestock manure management) and adaptation to climate change (for example, through more resilient crops). However, to date most countries have paid limited attention to post–farm gate activities that generate GHGs (such as energy use efficiency and food and plastic waste associated with processing, logistics, and consumption). Consumers are, however, increasingly conscious of the climate change effects of the food they consume.

FOOD SAFETY AND PERSISTENT NUTRITION PROBLEMS AS NEW SOURCES OF FOOD INSECURITY

In an increasingly urbanized East Asia region, concerns about food safety reflect the growing strains on the region's agri-food systems. The available evidence points to rising exposure of populations to food safety hazards, a significant incidence of foodborne illness[6] (table 2.3), and deepening consumer concerns about the contamination of local foods and the adequacy of prevailing governance structures to manage emerging risks. The region's food safety problems are diverse, stemming from environmental hazards (for example, contaminated soil and water); poor hygienic conditions and practices in farms, markets, food facilities, and households (bacterial and virus contamination); the improper use of fertilizers (high nitrate levels), pesticides, and antibiotics (exposure of farm workers, residues in products); and food and feed adulteration (adapted from World Bank 2019a). Food safety will become more significant in the future as a dominant share of consumers become reliant on purchased foods and ingredients and as diets continue to evolve beyond rice and other staple grains.

Zoonoses originating from livestock and wildlife represent a significant public health threat, and combatting their emergence is a public health and economic priority. Deforestation and habitat degradation and fragmentation, together with encroachment of livestock production into wild animal habitats, have enhanced zoonotic disease transmission through increased livestock-wildlife-human interface (Hassan 2014; Loh et al. 2015; McAleenan and Nicolle 2020). The frequency of emerging infectious diseases (EIDs) of livestock (for example, SARS/Avian flu and swine flu/H1N1) and particularly of wildlife origin (for example, MERS, Ebola, HPAI) is increasing exponentially[7] (Allen et al. 2017). In addition, the consumption of bushmeat and ecotourism have increased the possibilities of pathogen transmission, mutation, and spread (World Bank 2010). For instance, the inadequate regulation and poor hygiene and management of wet markets and wildlife trade likely further contributed to the recent outbreaks of SARS/Avian flu and COVID-19 (coronavirus) (McAleenan and Nicolle 2020; World Bank 2010).

TABLE 2.3 **Comparative public health burden: Disability-adjusted life years lost per 100,000 people, 2010**

COUNTRY	FOODBORNE DISEASE	TUBERCULOSIS	HIV/AIDS	HEPATITIS	MALARIA
China	272	173	146	30	1
Indonesia	693	1,728	479	108	255
Lao PDR	933	2,907	351	69	272
Malaysia	293	210	1,451	32	57
Myanmar	711	2,250	1,892	101	441
Philippines	293	1,338	21	66	61
Thailand	685	345	1,798	34	33
Vietnam	390	775	544	29	12
Global cases (millions)	33	40	—	—	66

Sources: World Health Organization Global Burden of Disease Statistics and Foodborne Disease Epidemiology Reference Group; Ekanayake, Ambrosio-Albala, and Jaffee 2019; World Bank 2019a.
Note: — = not available; HIV/AIDS = human immunodeficiency virus/acquired immune deficiency syndrome.

Costs of unsafe food and EIDs to public health and the economy amount to billions of dollars a year at the best of times. Foodborne diseases and pandemics such as the COVID-19 crisis can have a devastating effect on human capital through loss of life, livelihood, and nutrition, and interruption of essential health and education services. In East Asia and Pacific, food safety is also a competitiveness issue. The economic burden of unsafe food, in both absolute and relative terms, varies across countries according to the level of economic development and size of the population and the economy, being highest for China, Indonesia, and Thailand (World Bank 2019a). Productivity losses in developing East Asia attributed to unsafe food were estimated to total $45 billion in 2016 (figure 2.2). When considering the cost ($7 billion) of treating foodborne diseases, the domestic costs of unsafe food in emerging Asia may equal $70 billion, with China making up about $30 billion of this (World Bank 2019a). The countries of East Asia and emerging Asia have accounted for a disproportionate share of regulatory interceptions in industrial country markets attributable to food safety violations. A crude estimate of the trade-related costs would be equivalent to 3 percent of the value of high-value food exports, or just under $3 billion (Jaffee et al. 2019; World Bank 2019a). The global costs associated with the COVID-19 pandemic (precautionary and preventive measures, as well as economic and health costs) are likely to amount to trillions of dollars.[8]

Food safety concerns are compounded by persistent challenges with dietary quality. Throughout the region, poor or imbalanced diets contribute to high rates of child stunting,[9] widespread micronutrient deficiencies in both adult and child populations, and the rising incidence of overweight and obesity (FAO 2018). Environmental risks such as flood and drought can also cause shocks that result in increased incidence of poverty and malnutrition. The threat of

FIGURE 2.2

Estimated productivity loss due to foodborne disease, 2016

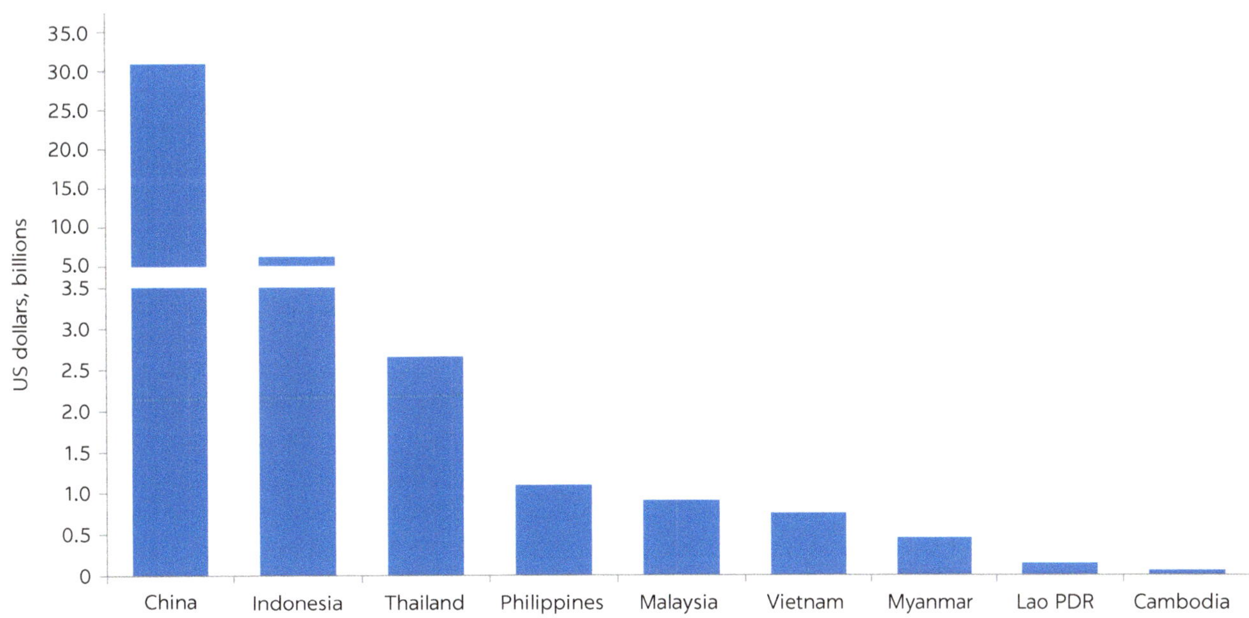

Source: Adapted from World Bank 2019b.

malnutrition from crop failures could increase with climate change (IFPRI 2019). The noncommunicable diseases (for example, cardiovascular diseases, diabetes) arising from overweight have also been identified as risk factors associated with the severity of COVID-19 infection (UN 2020). The region's success in addressing this "triple burden of malnutrition" will strongly enhance its human development progress as well as mitigate its future public health costs (World Bank 2019a). Moreover, investments in human capital have become more important as the nature of work has evolved. By improving health, in addition to skills, knowledge, and resilience, people can be more productive and innovative (World Bank 2018).

WEAKNESSES IN EAST ASIA'S AGRI-FOOD SYSTEMS EXPOSED BY COVID-19

The impacts of COVID-19 on the world's agri-food systems will last beyond the current crisis. The resilience of agri-food systems to future pandemics and external shocks, including climate change, needs to be improved. As governments rushed to minimize the movement and interaction of citizens to slow down transmission of the virus, the resulting restrictions together with the disease incidence resulted in disruptions in the food chain. Labor shortages at farms, the agri-food industry, logistics, and customs have constrained the movement and availability of food and agricultural production inputs and services.[10] Logistics disruptions of food and input supply chains have restricted the availability and access to food for consumers, which in some cases has led to local shortages and price hikes at a time when producers are experiencing declining farm gate prices and losses of food they are not able to sell (FAO 2020; PwC and FIA 2020). Lower incomes related to lockdowns have already resulted in shifts in consumer behavior and food consumption,[11] with repercussions for livestock production, food markets, and food loss and waste (*Economist* 2020; PwC and FIA 2020; World Bank 2020).

The COVID-19 crisis has underscored the urgency of addressing the region's existing food and nutrition challenges in a sustainable manner. Countries in East Asia and Pacific already face significant long-term food and nutrition security challenges from multiple demand- and supply-side constraints. The COVID-19 crisis could offer a window of opportunity for governments in the region to prioritize increasing the resilience of agri-food systems as part of the economic recovery. Taking advantage of this opportunity would, however, require governments to recognize the profound changes taking place in the supply of and demand for food in the region. The region's success in facing these unprecedented challenges relies on turning these changes into the foundations of a sustainable agri-food system.

On the demand side, the COVID-19 pandemic is reinforcing changes in consumer purchasing habits that were developing before the crisis. Evidence indicates the consumption of products of animal origin has declined because of fears associated with zoonotic contamination and the demand for plant-based proteins and immunity-enhancing foods (for example, vitamin-fortified food and drinks) has increased (IFPRI 2020; PwC and FIA 2020). The overall drive toward convenience and safety (and, in the COVID-19 context, containment) has stimulated greater demand for packaged foods, e-commerce, and more direct sourcing and delivery of food. Although some segments of the population may generate preferences for higher-quality fresh ingredients, others may stay with lower-cost products they relied on during the crisis (PwC and FIA 2020). The biggest impact is likely to be on the online food delivery market and the use of e-commerce platforms. For more details on e-commerce and nutrition, see chapter 5.

On the supply side, the COVID-19 pandemic has stimulated interest in more consolidated and traceable food chains, in urban farming (including in vertical farming), and in technologies that reduce reliance on labor. The disruptions in the food chain have given rise to an increase in sourcing from a more diverse supplier base as well as urban farming ventures. Such efforts can reduce the exposure of retailers and buyers to disruptions in individual geographic locations or products (that is, buyers engage with a larger number of smaller operations that produce a variety of products, or with local urban producers). Such engagement, however, requires building closer relationships with producers, especially smallholders, and building their capacity to produce (for example, through access to innovation, e-extension and other services, quality and safety standards, skills) and deliver at scale (for example, collective action, storage solutions, overcoming logistics bottlenecks, traceability). Labor shortages caused by disease incidence, lockdown conditions, and mass exodus of migrant laborers, as experienced in Malaysia and Thailand, have created interest in automation and e-services. Investments in "Industry 4.0" technologies such as robotics, artificial intelligence, and the Internet of Things to reduce dependence on on-site operational workers are likely to increase in response (for details, see chapter 5) (*Economist* 2020; PwC and FIA 2020). Greater demand for sustainable and safe livestock production that does not increase the risk of zoonotic outbreaks or of foodborne diseases also requires greater research and development efforts and innovation in parallel with improvements in veterinary services, animal health, and modern food safety regulations and enforcement mechanisms.

The disruptions in the food chain highlight the need for enhancing the resilience of agri-food systems, supported by a holistic approach to the prevention, control, and recovery of EIDs and zoonoses. Recognition of the interrelatedness of the respective health domains and of the risks that zoonotic diseases pose to public health has led to appeals for more horizontal interaction among the disciplines and the sector agencies, departments, and ministries that are responsible for public health, medical professions, veterinary services, and the environment. A crucial element is to prevent and minimize the local and global impact of epidemics and pandemics through a holistic food systems lens, integrating food and nutrition security dimensions intimately correlated to health issues. This undertaking requires the collaborative efforts of human, animal, and wildlife health experts; environmental experts; and food safety regulators, among others (WHO, n.d.; World Bank 2020).

Several international and national approaches to managing EIDs have generated useful lessons for East Asian countries. The approaches (One Health, among others), organized around the pillars of prevention, detection, response, and recovery, are summarized in table B.1 in appendix B. The lessons converge on several issues, including the range of human activity driving EIDs: hunting and the selling of wildlife in markets; keeping livestock in unsanitary conditions close to food preparation; co-location of multiple species in unnatural, captive conditions for sale; broader trading in wildlife or endangered species (even when farmed); indiscriminate land-use change in high-risk regions; indiscriminate use of antibiotics in agriculture; and eco-tourism (box 2.1) (McAleenan and Nicolle 2020; World Bank 2010). Other commonalities include the importance of early detection, clear communication and coordination, and governance by

BOX 2.1

The main drivers of emerging infectious diseases

The increase in animal numbers has led to a significant restructuring of how production is organized spatially, perhaps most notably in peri-urban areas, and particularly with respect to pig and poultry production. The scale of large commercial farms has increased dramatically and has become concentrated in relatively small areas. With improved transport, large farms tend to move away from large cities to areas with abundant feed supplies.

The capacity to protect livestock and manage diseases remains weak, especially among small-scale producers. Larger commercial producers can generally better afford to invest in more sophisticated forms of biosecurity than can small producers, who continue to operate with little if any biosecurity. Scant attention has been given to innovations that can help small producers meet their biosecurity needs in their resource-poor circumstances.

Inappropriate vaccination and drug use raise risks. The inadequacy of the health systems in the East Asia and Pacific Region causes gaps in vaccination coverage and suboptimal use of drugs, leading to drug resistance and hence increased risk of newly emerging pathogens. Adding antibiotics to livestock feed for nontherapeutic purposes is another cause of induced resistance to antibiotics in animal source foods.

Exploitative farming systems are conducive to flare-ups and persistence of existing agents. Poor working conditions and animal housing conditions can lead to hazardous interactions between livestock and humans, and between livestock and wild species, setting the stage for the flare-up of novel agents and for the persistence of existing agents. Today, these hot spots for emerging infectious diseases are increasingly connected to the larger world through legal and illegal trade and human traffic in a context of globalization.

A diverse reservoir of influenza viruses also circulates in wild birds, and contacts between these birds and domestic poultry and pigs are common. These contacts lead to human exposure and to the exchange of viruses and genetic material between humans and animals.

Major land-use changes, including intensification and deforestation, increase the wildlife-livestock-human interface and with it the risk of novel pathogens jumping to humans. The rapidly growing livestock sector has been a principal driver in the conversion of natural habitats into pastures and cropland. The intensification of agriculture with increased use of inorganic fertilizer, together with increasing livestock density, has been a major source of water pollution, and often provides favorable environments in which novel pathogens emerge. Major land-use changes degrade ecosystems, fragment habitats, and change host-pathogen dynamics. Degraded ecosystems with diminished biodiversity tend to favor opportunistic or generalist species, many of which are disease reservoirs.

Hunting, poaching, and bushmeat trade, and both legal and illegal trade in live animals have increased rapidly over the past few decades and are a major factor in the spread of diseases. In addition, increased eco-tourism and activity near wildlife areas have increased accidental transmission of pathogens to humans.

Changes in long-term and seasonal weather patterns will have major effects on disease behavior such as spreading patterns, diffusion range, and introduction and persistence in new habitats. The extension of vector habitats will be a major factor in the impact of climate change on the spread of infectious diseases.

Sources: Hassan 2014; McAleenan and Nicolle 2020; World Bank 2010.

independent dedicated organizations, and the need for multidisciplinary teams for prevention, joint preparedness and priority setting, and implementation of EID interventions. However, even though prevention is widely discussed, it is rarely addressed in practice using the above approaches. Apart from the lack of prevention of EIDs, the recovery interventions are also often missing, which may contribute to persistent risky human activity in "hot spots" and the reemergence of existing EIDs.

NOTES

1. For example, in Vietnam crop productivity has fallen gradually over the years. In the animal husbandry and aquaculture sector, diseases have become widespread and seriously affected both the productivity and income of farmers (FFTCP-AP 2015).
2. Lack of knowledge leads to excessive use, inappropriate timing, poor choice of pesticide, and limited awareness of alternative methods.
3. China, Indonesia, the Philippines, Thailand, and Vietnam produce half of all plastic waste in the world's oceans. Plastic is not properly disposed of and ends up in rivers and oceans (World Bank 2019b). For details on plastic management in developing East Asia, see appendix A.
4. The Intergovernmental Panel on Climate Change notes that infectious diseases will likely increase because of a warmer climate. Vector-borne diseases (those that spread via unaffected third-party species, especially parasites) are likely to spread to new regions (IPCC 2018).
5. Agriculture generates 21–24 percent of total GHG emissions from multiple sources, such as burning of debris, overexploitation of peatlands, rice, and livestock production (Acosta et al. 2019). Methane is released from excessive nitrogen fertilizer use, flooded rice, and ruminant livestock production. Carbon dioxide emissions come from forest clearing and degradation, particularly of tropical peatlands (Posa, Wijedasa, and Corlett 2011; Scherr et al. 2015).
6. Animal products account for 50 percent or more of the burden of foodborne diseases in Cambodia, China, Lao PDR, and the Philippines. This share is below 25 percent, perhaps because of dietary restrictions (Indonesia and Malaysia) or other reasons for low per capita consumption of some animal products (Myanmar). Contamination in processed foods is generally not a major source of foodborne diseases (World Bank 2019a).
7. About five new EIDs and about three new zoonoses occur annually (Allen et al. 2017). Antimicrobial resistance in human pathogens is another major public health threat, partly affected by the use of antibiotics in livestock production (WHO, n.d.).
8. An annual investment of $3.4 billion in veterinary systems could avert $6.7 billion losses incurred through delayed or inadequate responses to zoonoses (World Bank 2020). The expected annual benefit of pandemic prevention is greater than $30 billion (World Bank 2012).
9. Stunting rates (children under age five) in Cambodia, Indonesia, and Myanmar are close to 30 percent and in Vietnam 25 percent (FAO 2018).
10. Countries have addressed these by, for example, allowing certain relaxations for movement of food and inputs and through promotion of urban farming, collective centers, e-services, e-commerce for agriculture, and loans and support to agribusiness (IFPRI 2020; PwC and FIA 2020).
11. Shifts in consumer behavior include choosing lower-cost food items; avoidance of food markets and animal products; greater demand for plant-based proteins, e-commerce, (plastic) packaging, and local food; panic buying; and home cooking rather than restaurant eating.

REFERENCES

Acosta, M. H., P. Thornton, D. Mason-D'Croz, and J. Palmer. 2019. "Transforming Food Systems under a Changing Climate: Changing Diets and Transforming Food Systems: Key

Messages." October. CGIAR Research Program on Climate Change, Agriculture, and Food Security.

Allen, T., K. A. Murray, C. Zambrana-Torrelio, S. S. Morse, C. Rondinini, M. Di Marco, N. Breit, K. J. Olivial, and P. Daszak. 2017. "Global Hotspots and Correlates of Emerging Zoonotic Diseases." *Nature Communications* 8, article number 1124. https://www.nature.com/articles/s41467-017-00923-8.

Economist. 2020. "How to Feed the Planet. The Global Food Supply Chain Is Passing a Severe Test. Keeping the World Fed." May 9, 2020. https://www.economist.com/leaders/2020/05/09/the-global-food-supply-chain-is-passing-a-severe-test.

EJF (Environmental Justice Foundation). 2003. "What's Your Poison? Health Threats Posed by Pesticides in Developing Countries." Environmental Justice Foundation, London.

Ekanayake, I. J., M. Ambrosio-Albala, and S. M. Jaffee. 2019. "From Rice to RICE: Integrated Strategies for Myanmar to Realize a More Resilient, Inclusive, Competitive and Environmentally Sensitive Food System." Presentation summarizing the "Benchmarking of Myanmar's Food Systems" report, World Bank, Washington, DC.

FAO (Food and Agriculture Organization of the United Nations). 2015. *Climate Change and Food Systems: Global Assessments and Implications for Food Security and Trade*. Rome: FAO.

FAO (Food and Agriculture Organization of the United Nations). 2018. *Dynamic Development, Shifting Demographics, and Changing Diets*. Bangkok: FAO.

FAO (Food and Agriculture Organization of the United Nations). 2020. "Addressing the Impacts of COVID-19 in Food Crises|April–December 2020: FAO's Component of the Global COVID-19 Humanitarian Response Plan." FAO, Rome. https://doi.org/10.4060/ca8497en.

FFTCP-AP (Food and Fertilizer Technology Center for the Asian and Pacific Region). 2015. "Vietnam's Policies on Agricultural Restructuring." Food and Fertilizer Technology Center, Taipei.

Fuglie, K., M. Gautam, A. Goyal, and W. F. Maloney. 2020. *Harvesting Prosperity: Technology and Productivity Growth in Agriculture*. Washington, DC: World Bank.

Hassan, L. 2014. "Emerging Zoonoses in Domesticated Livestock of Southeast Asia." *Encyclopedia of Agriculture and Food Systems* 2014: 68–81.

IFPRI (International Food Policy Research Institute). 2019. *Policies to Support Investment Requirements of Indonesia's Food and Agriculture Development during 2020–2045*. Manila: Asian Development Bank.

IFPRI (International Food Policy Research Institute). 2020. "Getting Biofortified Food on Everyone's Plate: HarvestPlus 2019 Annual Report." IFPRI, Washington, DC. https://www.ifpri.org/publication/getting-biofortified-food-everyones-plate-harvestplus-2019-annual-report.

IPCC (Intergovernmental Panel on Climate Change). 2018. *Global Warming of 1.5°C*. Special Report. Geneva: IPCC. http://www.ipcc.ch/report/sr15/.

Jaffee, S., S. Henson, L. Unnevehr, D. Grace, and E. Cassou. 2019. "The Safe Food Imperative: Accelerating Progress in Low- and Middle-Income Countries." World Bank, Washington, DC.

Loh, E. H., C. Zambrana-Torrelio, K. J. Olivial, T. L. Boqich, C. K. Johnson, J. A. Mazet, W. Karesh, and P. Daszak. 2015. "Targeting Transmission Pathways for Emerging Zoonotic Disease Surveillance and Control." *Vector Borne Zoonotic Diseases* 15 (7): 432–37. https://www.ncbi.nlm.nih.gov/pubmed/26186515.

McAleenan, B., and W. Nicolle. 2020. "Outbreaks and Spillovers: How the UK and International Community Can Lower the Risks of Zoonotic Disease." Policy Exchange 2020, London.

Posa, M. R., L. S. Wijedasa, and R. T. Corlett. 2011. "Biodiversity and Conservation of Tropical Peat Swamp Forests." *BioScience* 61 (1): 49–57.

PwC (PricewaterhouseCoopers) and FIA (Food Industry Asia). 2020. "Maintaining Food Resilience in a Time of Uncertainty. Understanding the Importance of Food Value Chains in ASEAN and How to Ensure Their Resilience during the COVID-19 Crisis." PricewaterhouseCoopers. Commissioned by Food Industry Asia.

Sapkota, A., M. Kucharski, J. Burke, S. McKenzie, P. Walker, and R. Lawrence. 2008. "Aquaculture Practices and Potential Human Health Risks: Current Knowledge and Future Priorities." *Environment International* 34 (8): 1215–26.

Scherr, Sara J., Kedar Mankad, Steven Jaffee, and Christine Negra, with Tanja Havemann, Janjarang Kijtikhun, Uray Endang Kusumjaya, Samiksha Nair, and Naomi Rosenthal. 2015. *Steps toward Green: Policy Responses to the Environmental Footprint of Commodity Agriculture in East and Southeast Asia*. Washington, DC: EcoAgriculture Partners and the World Bank.

Searchinger, T. 2016. "Possibilities for Modernization of Agriculture and Mitigation of Its Greenhouse Gas Emissions in China." Short Discussion Memo of Possible Areas of World Bank/China Collaboration. Unpublished.

UN (United Nations). 2020. "Policy Brief: A Disability-Inclusive Response to COVID-19." United Nations, New York. https://www.who.int/who-documents-detail/policy-brief-of-the-un-secretary-general-a-disability-inclusive-response-to-covid-19.

WHO (World Health Organization). n.d. "Managing Public Health Risks at the Human-Animal-Environment Interface." World Health Organization, Geneva. https://www.who.int/zoonoses/en/.

World Bank. 2004. "The Socialist Republic of Vietnam: Coffee Sector Report." Report 29358-VN, World Bank, Washington, DC.

World Bank. 2010. *People, Pathogens, and Our Planet. Volume 1: Towards a One Health Approach for Controlling Zoonotic Diseases*. Washington, DC: World Bank.

World Bank. 2012. *People, Pathogens and Our Planet: Volume II. The Economics of One Health*. Washington, DC: World Bank.

World Bank. 2017. "Megatrends and the Trajectory of Agro-food Systems in East Asia." Agriculture and Food Forum, Washington, DC, March 6–7.

World Bank. 2018. "The Human Capital Project." In *World Development Report 2019: The Changing Nature of Work*. Washington, DC: World Bank. doi:10.1596/978-1-4648-1328-3.

World Bank. 2019a. *Managing Headwinds. East Asia and Pacific Economic Update* (April). Washington, DC: World Bank. doi:10.1596/978-1-4648-1412-9.

World Bank. 2019b. "Marine Plastic Debris. A Regional Engagement Framework for Action for the World Bank East Asia and Pacific Region." Unpublished manuscript, World Bank, Washington, DC.

World Bank. 2020. "COVID-19: Opportunities to Strengthen China's Animal Disease Prevention and Food Safety Strategies for Transforming One Health Outcomes. Unpublished. World Bank, Washington, DC.

WRI (World Resources Institute). 2013. *Creating a Sustainable Food Future: Interim Findings. A Menu of Solutions to Sustainably Feed More than 9 Billion People by 2050*. Washington, DC: World Resources Institute.

3 The Vital Role of Innovation in Agricultural Transformation

EAST ASIAN AGRI-FOOD SYSTEMS NEED TO EMBRACE INNOVATIONS THAT FOSTER PRODUCTIVITY, SUSTAINABILITY, AND HEALTH

Amid the pressures on the agri-food system, developing East Asian countries also share the desire to move up the economic ladder. The countries can be categorized as agrarian, transition, or urban-mature on the basis of agriculture sector characteristics, development, and the sector's importance for the overall economy (World Bank 2007). The three categories are well aligned with low-income country, middle-income country (MIC), and high-income country status. Table 3.1 summarizes the three categories and identifies East Asian countries that fall into each category and subcategory (upper, mid-range, and lower MIC).[1]

Rapid income-level transitions are, however, challenging because they require rethinking the development model and simultaneously investing in old and new economic capabilities.[2] Economies that have successfully made the transition from middle-income to high-income status (for example, the Republic of Korea; Taiwan, China; and Singapore) share several common characteristics: good governance, competition, labor mobility, innovation, integration into the global economy, financial development, investment in human capital, and sustainable infrastructure. Recent studies show that returns on innovation are highest in MICs[3] (EBRD 2018). Although the returns to innovation are often positive and high, they may become negative with distance from the frontier, that is, when other conditions, such as existing capacity (public and private), scientific infrastructure, access to other inputs, cost of doing business, trade regime, intellectual property rights, and capital markets, limit the returns to innovation investment (Cirera and Maloney 2017; Goni and Maloney 2014). More than ever, country transition relies on moving to a knowledge-based economy, fueled by innovation and investment in research and development (R&D), and complemented with other necessary conditions and investments.

Harnessing science and innovation will be critical for agri-food systems to make the transition toward resilience and sustainable growth. The region has witnessed impressive gains from past and current innovations, including

TABLE 3.1 **Country categorization according to agriculture sector characteristics and importance**

AGRICULTURE CATEGORY	AGRICULTURE SECTOR	COUNTRY INCOME STATUS	COUNTRY EXAMPLES
Urban-Mature	Most of the country's population is in urban areas, and the agricultural sector is well integrated into global markets. Poverty is principally an urban problem. Many countries in Latin America and the Caribbean, Europe, and Central Asia are in this category.	HIC	Australia; Japan; Korea, Rep.; Netherlands
Transition	Mostly countries in Asia, where economic growth is now led by the industrial sector and the economy is rapidly urbanizing. Growing urban demand, especially for higher-value products such as livestock and horticultural crops, is resulting in structural shifts in the agricultural economy. Even so, large areas of the rural economy still have high poverty rates and are not integrated into the growth process. Many countries in South Asia, East Asia and Pacific, the Middle East, Latin America, and North Africa are in this category, including China and India.	MIC Upper MIC Middle MIC Lower MIC	 Chile, China, Malaysia, Thailand Indonesia, Philippines, Vietnam Cambodia, Lao PDR, Myanmar
Agrarian	Usually relatively small, mostly Sub-Saharan African countries that depend primarily on staple food crop production and rely on agriculture for their economic growth. The agricultural markets are not well integrated, transport and logistics are costly, and private investment in rural areas is still limited.	LIC	Malawi, Nepal, Afghanistan

Sources: Original table for this publication, drawing on World Bank 2007 for agriculture sector categories; World Bank Databank, 2020, for country examples in last column (https://databank.worldbank.org/home.aspx).
Note: HIC = high-income country; LIC = low-income country; MIC = middle-income country.

technological, management, organizational, and business model innovations (chapter 4). To date, much of the attention on agricultural innovation has been directed toward the innovation and adoption processes taking place at the farm gate—varieties, breeds, sustainable management practices, and input use—and the transfer-adoption process among farmers, farmer groups, and agro-enterprises. The role of the public sector has been prominent in on-farm innovations, whereas the private sector has gradually expanded from innovation in diverse inputs, delivery systems, and information and communication technology–based services to food and beverages. The scale of the current challenges means, however, that wider adoption of these innovations, and continued improvements to them, along with strengthening innovation capacity, will be necessary to offset the growing challenges. The inclusion of the millions of smallholders, among them women and poor or marginalized people, in the transformation of the sector will also be essential. The agricultural innovation process and the factors affecting it are discussed in box 3.1.

Government science, technology, and innovation (STI) policies in most developing East Asian countries have traditionally focused on food security and agriculture productivity, that is, production aspects, while giving cursory attention to wider agri-food system needs. STI policies and investments must expand the focus beyond food security and production and accommodate the needs of the entire agri-food system from farm to fork. Much greater attention should be paid to sustainability, resilience, and post–farm gate innovations: in the modern agri-food system most of the value is generated from post–farm gate activities (Reardon and Timmer 2014). This is also where private sector innovation will increasingly focus. To date, the role of the public sector in food processing has been limited (rightly so) and mostly focused on regulatory aspects, intellectual property rights, and the wider enabling environment.

BOX 3.1

Factors affecting the agricultural innovation process

Agricultural productivity growth and agri-food system transformation rely heavily on research, technology, and innovation (Hall and Dijkman 2019). The role of research and development (R&D) has been the most influential factor in growth—returns to public R&D investments have been very high, about 49.5 percent in the East Asia and Pacific Region (Alston et al. 2000). The growth in private agricultural R&D across multiple industrial sectors also suggests that firms have found it profitable to invest in agricultural research. In developed Organisation for Economic Co-operation and Development countries, private agriculture R&D has stabilized at about 50 percent of total agricultural R&D. By late in the first decade of the 2000s, the private sector accounted for about 25 percent and 10 percent of total agricultural R&D spending in China and Vietnam, respectively. Although private R&D was channeled to fertilizers, pesticides, seed and biotechnology, and machinery in China, by far most private R&D (about 50 percent) was channeled to food manufacturing and plantations (Fuglie et al. 2020).

Agricultural innovation is, however, more than agricultural R&D and adoption of agricultural technology. Innovation refers to the process of creating and putting into use combinations of knowledge from many different sources. Innovation may be brand new, such as invention, but usually it involves new combinations of existing knowledge. The use of this knowledge must be novel to a farmer or a firm and their neighbors and competitors, but not necessarily new globally (World Bank 2006a). An innovation system, by contrast, is a network of organizations, enterprises, and individuals focused on bringing new products, new processes, and new forms of organization into economic use, together with the institutions and policies that affect their behavior and performance (World Bank 2006a).

Evidence from agricultural innovation studies confirms that the ability to innovate depends on technical skills and R&D and is closely related to interaction among relevant public and private actors. Such interaction relates to, for example, collective action (organizational innovation), knowledge exchange, skills, and incentives and resources for collaboration and is affected by conditions (such as regulations, intellectual property rights, finance, infrastructure) that enable adoption and innovation (World Bank 2012). Multiple documented examples of innovation processes in agri-food systems are available (Kelly 2019; World Bank 2006a, 2006b, 2012). Thus, innovation may arise spontaneously from the interaction of different actors but may significantly benefit from, for example, incentives and support for partnerships and a favorable enabling environment.

Notable Asia-specific examples of agricultural innovation include stimulation of small farm machinery development in China in response to widespread adoption of the Green Revolution technology packages; development of the cassava starch industry in Thailand; transformation of the Thai poultry sector into the world's leading poultry exporter along with the challenges and innovation processes for addressing the food safety crisis of the poultry sector; and the development of the A-vitamin-rich Golden Rice variety in a collaboration between public (International Rice Research Institute, national agricultural research systems) and private R&D and the challenges with its adoption in a complex socioeconomic policy environment (Huang, Rozelle, and Hu 2007; Kelly 2019).

THE AGRICULTURAL INNOVATION SYSTEMS APPROACH

Agricultural innovation takes place within an agricultural innovation system (AIS), principally within a network of relevant public and private actors affected by a wider policy, governance, and enabling environment. There are many schematic presentations of AIS; this report features one adapted from Fuglie et al. 2020. Figure 3.1 illustrates the AIS in a way that systematically maps out the factors affecting innovation. It treats innovation as the accumulation of knowledge, labor, and physical capital. The figure highlights that farms, farmer organizations, and firms are the critical players in the system and that their decisions about accumulating capital, labor, or knowledge need to be jointly considered. Figure 3.1 broadly distinguishes the demand for factors from the supply to highlight that, without demand from farms, farmer organizations, firms, and the wider economy, supply-side policies to generate or disseminate relevant knowledge are equivalent to pushing on a string. The division between the two sets of variables is not sharp, particularly in the knowledge area, and the bi-directional arrows aim to capture the feedback relationship between farms and firms on the one hand and knowledge institutions on the other (Fuglie et al. 2020). Coordinators and knowledge brokers are shown in the middle given that they may support interaction and innovation processes among supply- and demand-side actors. Appendix C describes the supply of and demand for, as well as the enabling environment for or barriers to, innovation.

Given limited resources, each country needs to assess the most binding constraints to innovation. Both coordination and governance (not captured by figure 3.1) play crucial roles in guiding the innovation process and the associated investments in strategic areas, that is, priorities to address or remedy to foster innovation. Coordination may facilitate demand articulation, co-innovation among diverse actors, and alignment of other essential investments for innovation to take hold. Innovation governance refers to the systems and practices for setting priorities and agendas, designing and implementing policies, and obtaining knowledge about their impacts (World Bank 2012).

The maturity of the AIS in any given country is typically aligned with the state of its agri-food system and overall economic development. Figure 3.2 provides an

FIGURE 3.1

Schematic presentation of agricultural innovation system

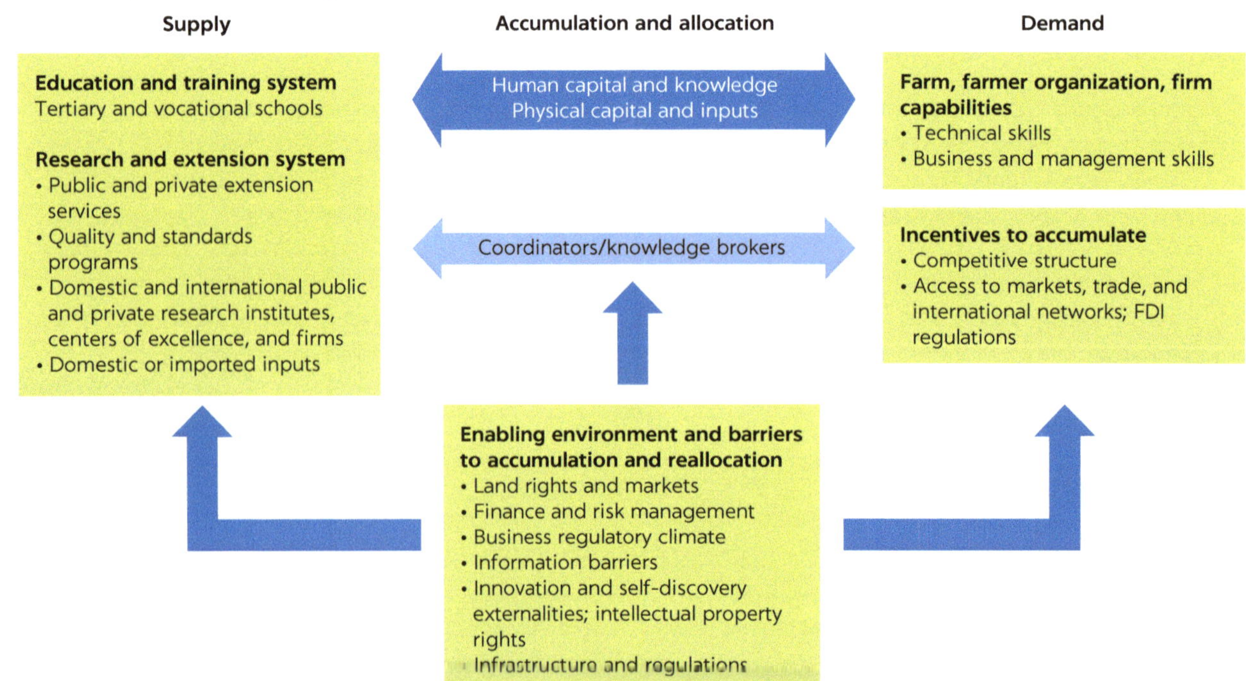

Source: Adapted from Fuglie et al. 2020.
Note: FDI = foreign direct investment.

overview of an AIS. The trends and approaches columns refer to a range of capacity and practices embodied in each AIS element. These AIS elements typically become more sophisticated as the country's agriculture sector develops (for example, from agrarian to urban-mature) and with overall economic development (from low-income country to high-income country). The main AIS elements are discussed in greater detail in chapter 6.

The AIS approach can bring attention to inclusive systemwide innovation and add value to existing efforts to reform agri-food systems. To date, the lack of a holistic, inclusive, and proactive approach across relevant disciplines and organizations has often prevented policy makers and investors from making timely

FIGURE 3.2

The maturity of an agricultural innovation system varies with the agri-food sector and the overall economy

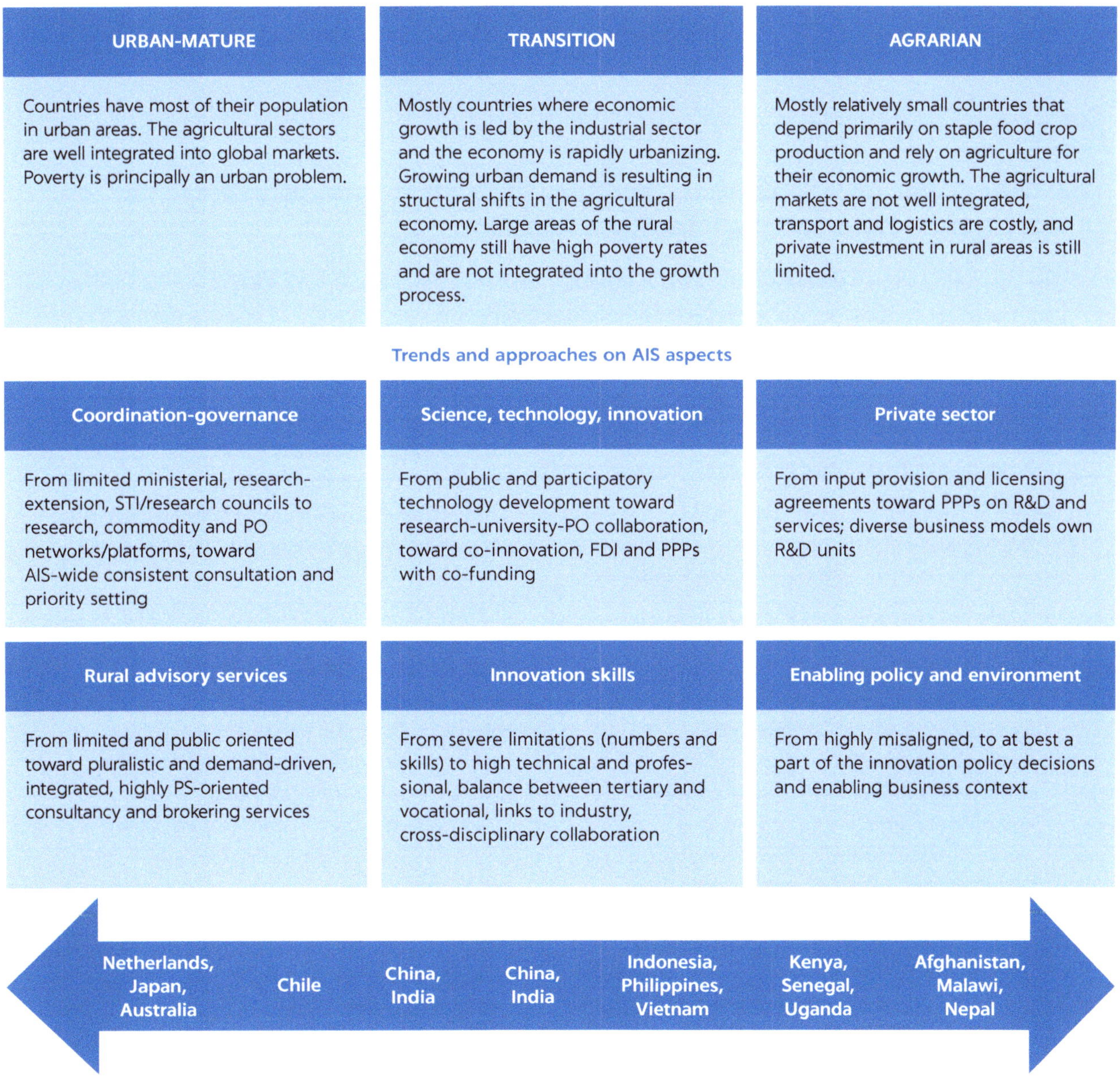

Source: Original work for this publication.
Note: AIS = agricultural innovation system; FDI = foreign direct investment; PO = producer organization; PPP = public-private partnership; PS = private sector; R&D = research and development; STI = science, technology, and innovation.

decisions to improve agri-food systems' ability to address existing and oncoming risks (for example, with plastic use, zoonoses, weather) and impacts and to take advantage of new opportunities. Shaping agri-food systems to deliver improved sustainability and human and animal health outcomes requires a combination of improved knowledge (for example, understanding of how food systems, nutrition, and the environment interact, and analysis of options and trade-offs), coordination, and sound policies, regulations, and investments across the agri-food system (adapted from IFPRI 2020; Townsend et al. 2016; WEF 2018). The AIS approach can help strategic leaders and stakeholders identify the key drivers, bottlenecks (for example, capacity, technology, enabling environment), and resource requirements and help them develop an impact-oriented action plan for a given issue.

The AIS is a natural fit with the multidisciplinary emerging infectious disease (EID) management approaches. AIS, with its focus on the innovation process, can support development of interventions that target prevention of and recovery

BOX 3.2

Potential entry points for the agricultural innovation systems approach in management of emerging infectious diseases

The agricultural innovation systems (AIS) approach can contribute to interventions targeting prevention of and recovery from emerging infectious diseases (EIDs). The AIS approach is holistic and a natural fit with the multidisciplinary EID management approaches.

Prevention of EIDs. AIS can be applied to address, for instance,

- Development of sustainable livestock production practices (such as sanitary conditions, antibiotics use, and smallholders' biosecurity needs; vaccine and drug development and services in hard-to-reach locations; contained production conditions; efficient and sustainable feed and feeding practices; and animal trade);
- Land use in agriculture (such as efficient and sustainable production practices reducing pollution and pressure on new land);
- Alternatives to livestock production and wildlife poaching (such as other more sustainable production and livelihood sources);
- Enhanced traceability systems for livestock products; and
- Incentives and disincentives to market failures that result in continuation of unsustainable practices

Response to and recovery from EIDs. The AIS approach can be used to ramp up the use of existing feasible innovations, such as

- E-services (targeting producers and firms in lockdown conditions);
- Precision agriculture and automation (supporting logistics and labor shortages);
- Traceability and e-commerce systems (matching demand with supply, controlling food loss and waste);
- Production of alternative protein sources (as a response to avoidance of products of animal origin) and nutritious foods (for example, biofortification, fortification);
- Safety standards and practices in local markets;
- Collective action (for example, for trade and logistics, services and input supply); and
- Sustainable production practices and input supply (for example, improved seed, biofortified seed).

These interventions are not simple fixes: they require coordination and significant collaboration between public and private actors and the presence of incentives to key stakeholders.

Source: World Bank.

from EIDs (box 3.2), which are the two areas largely missing in existing EID management interventions (appendix C) (Carlin et al. 2019; McAleenan and Nicolle 2020). A holistic approach on its own is not sufficient—it needs to be complemented with leadership, coordination, capacity, incentives, and resources.

The ongoing agri-food system transformation process is influenced by dynamic innovation in the broader economy. Despite the pressing challenges of environmental sustainability and public health, agricultural productivity remains one of the primary objectives in the transformation of agri-food systems over the coming decades. Agricultural producers are already adopting a wide array of innovations and technologies, ranging from more basic technologies and innovations at the farm level (improved crops and livestock breeds, irrigation, e-services) to disruptive technologies across the entire value chain (such as vertical farming, precision agriculture, genetically modified crops, data systems) (chapter 4). The agri-food system is, however, influenced by innovation in the broader economy, such as digitalization, artificial intelligence, big data systems, and biotechnology, as well as nutrition and manufacturing (chapter 5). To date, the agri-food sector overall has been slow to harness the power of these innovations (OECD 2011; WEF 2018). It needs to catch up, in both innovation and investments to address the challenges it faces and to reap the opportunities while making sure that new innovations and production techniques are accepted by society. The AIS of developing East Asian countries (chapter 6) indicates a readiness to develop domestic innovation capacity. With the right combination of policies and instruments, East Asian countries could move toward more resilient and sustainable food systems.

NOTES

1. MICs are further defined as lower-middle-income and upper-middle-income countries, with gross national income per capita of $1,006–$3,955 and $3,956–$12,235, respectively (in 2018). MICs are home to 75 percent of the world's population, 62 percent of the world's poor, and about one-third of global GDP, and are major engines of global growth. For this report, China, Malaysia, and Thailand are designated "upper MICs"; Indonesia, the Philippines, and Vietnam are designated "mid-range MICs"; and Cambodia, Lao PDR, and Myanmar are designated "lower MICs."
2. The upper MICs may be at risk of falling into the so-called middle-income trap, that is, they can no longer compete against lower-cost neighboring countries in less skill-intensive activities, but they still lack the science, technology, and innovation capabilities and skills needed to move into higher-value-added activities and compete with more industrial neighbors (UNCTAD 2018).
3. In high-income countries, returns on R&D are lower because production is already technologically more advanced and subject to the law of diminishing returns. In poorer countries, returns are constrained by the scarcity of necessary skills and a lack of scientific infrastructure, as well as by other institutional weaknesses affecting the ability of innovative firms to grow and access export markets (EBRD 2018).

REFERENCES

Alston, J. M., C. Chan-Kang, M. Marra, P. G. Pardey, and T. Wyatt. 2000. "A Meta-Analysis of Rates of Returns to Agricultural R&D." International Food Policy Research Institute, Washington, DC. http://www.ifpri.org/sites/default/files/publications/rr113.pdf.

Carlin, E., C. Machalaba, F. Berthe, K. Long, and W. Karesh. 2019. *Building Resilience to Biothreats*. New York: EcoHealth Alliance. 10.13140/RG.2.2.26016.64001.

Cirera, X., and W. F. Maloney. 2017. *The Innovation Paradox: Developing-Country Capabilities and the Unrealized Promise of Technological Catch-Up*. Washington, DC: World Bank. doi:10.1596/978-1-4648-1160-9.

EBRD (European Bank for Reconstruction and Development). 2018. *Eight Things You Should Know about Middle-Income Transitions*. London: EBRD.

Fuglie, K., M. Gautam, A. Goyal, and W. F. Maloney. 2020. *Harvesting Prosperity: Technology and Productivity Growth in Agriculture*. Washington, DC: World Bank.

Goni, E., and W. F. Maloney. 2014. "Why Don't Poor Countries Do R&D?" Policy Research Working Paper 6811, World Bank, Washington, DC.

Hall, A., and J. Dijkman. 2019. *Public Agricultural Research in an Era of Transformation: The Challenge of Agri-Food System Innovation*. Rome and Canberra: CGIAR Independent Science and Partnership Council Secretariat and Commonwealth Scientific and Industrial Research Organisation.

Huang, Q. Q., S. Rozelle, and D. Hu. 2007. "Pump-Set Clusters in China: Explaining the Organization of the Industry That Revolutionized Asian Agriculture." *Asia-Pacific Development Journal* 14 (December): 75–105.

IFPRI (International Food Policy Research Institute). 2020. *IFPRI 2020 Global Food Policy Report. Building Inclusive Food Systems*. Washington, DC: International Food Policy Research Institute. http://ebrary.ifpri.org/utils/getfile/collection/p15738coll2/id/133646/filename/133857.pdf.

Kelly, J. ed. 2019. *Public Agricultural Research and Development in an Era of Transformation: The Challenge of Agri-Food System Innovation. Resource Document I: Case Studies*. Rome and Canberra: CGIAR Independent Science and Partnership Council Secretariat and Commonwealth Scientific and Industrial Research Organisation.

McAleenan, B., and W. Nicolle. 2020. "Outbreaks and Spillovers. How the UK and International Community Can Lower the Risks of Zoonotic Disease." Policy Exchange 2020, London.

OECD (Organisation for Economic Co-operation and Development). 2011. "ISIC REV. 3. Technology Intensity Definition, 2011." OECD, Paris. https://www.oecd.org/sti/ind/48350231.pdf.

Reardon, T., and P. Timmer. 2014. "Five Inter-Linked Transformations in the Asian Agrifood Economy: Food Security Implications." *Global Food Security* 3 (2): 108–17. https://doi.org/10.1016/j.gfs.2014.02.001.

Townsend, R., S. Jaffee, M. Hoberg, Y. Tanimichi, A. M. Htenas, M. Shekar, H. Ziauddin, G. Madhur, H. Kray, L. Ronchi, H. Sarwat, L. K. Elder, and D. M. Eugene. 2016. *Future of Food: Shaping the Global Food System to Deliver Improved Nutrition and Health*. Washington, DC: World Bank Group. http://documents.worldbank.org/curated/en/474831468186561685/Future-of-food-shaping-the-global-food-system-to-deliver-improved-nutrition-and-health.

UNCTAD (United Nations Conference on Trade and Development). 2018. *Science, Technology & Innovation Policy Review: Thailand*. Geneva: United Nations.

WEF (World Economic Forum). 2018. "Innovation with a Purpose: The Role of Technology Innovation in Accelerating Food Systems Transformation. System Initiative on Shaping the Future of Food Security and Agriculture." WEF and McKinsey & Company, Geneva.

World Bank. 2006a. "Enhancing Agricultural Innovation: How to Go Beyond the Strengthening of Research Systems." World Bank, Washington, DC.

World Bank. 2006b. "Technology, Adaptation, and Exports: How Some Developing Countries Got It Right." World Bank, Washington, DC.

World Bank. 2007. *World Development Report 2008: Agriculture for Development*. Washington, DC. World Bank. https://openknowledge.worldbank.org/handle/10986/5990. License: CC BY 3.0 IGO.

World Bank. 2012. *Agricultural Innovation Systems. Investment Sourcebook*. Washington, DC: World Bank.

4 Past Agricultural Innovation in Perspective

UNDISPUTED SUCCESS OF PAST AGRICULTURE PRODUCTIVITY AND FOOD SECURITY ACHIEVEMENTS

Food security and poverty reduction goals have largely driven the investments in agricultural innovation over the past four decades. In the 1960s, the world's population was soaring, extreme poverty was widespread, agricultural productivity (particularly in food crops) was very low, and land for agriculture was limited in many areas (Fuglie et al. 2020; Gollin 2010; Timmer 2002).

The Green Revolution of the 1960s and 1970s saw the spread of new modern varieties (MVs), fertilizers, and other agricultural technologies (pesticides, machinery) to smallholder farmers. Adoption was enabled by extension services and wider economic reforms (Hazell 2009) (table D.1 in appendix D). The MVs were by far the most significant innovations, developed initially by international agricultural research centers in partnership with national agricultural research systems, drawing on existing germplasm in the Asia region. The Green Revolution was a continuing process of change rather than a single event (Evenson and Gollin 2003; Hazell 2009). Appendix D discusses the shifting priorities: the first-generation MVs in the 1960s (irrigated areas, rice and maize), the MVs in the 1970s (rainfed areas, other crops), and MVs from 1981 to 2000 (pest resistance, productivity).

The response of producers and the agriculture sector to Green Revolution innovations has been impressive—crop yields have increased nearly sixfold.[1] By 2000, MVs of rice, maize, and wheat had been adopted on more than 80 percent of the crop area in East Asia (table 4.1). Irrigation, fertilizer, and pesticide use have also increased significantly over the years. Pesticide use has since plateaued at 2010 levels, including in China, Malaysia, and Vietnam, which largely drove pesticide use (box D.2 in appendix D). Table 4.2 presents past and current rice and maize yields and overall use of inputs (excluding MVs, labor, and machinery)[2] in each country. Overall, yields tend to be higher in countries with greater input use and availability of irrigation.

TABLE 4.1 **Key innovations driving the initial production miracle in East Asia, 1960–2015**

	CROP			OTHER TECHNOLOGIES					
	RICE	MAIZE	WHEAT[a]	IRRIGATION	FERTILIZER	TRACTORS	PESTICIDE		
								TONS PER YEAR	
YEAR	AREA IN MODERN VARIETIES PER CROP HARVESTED AREA (%)			SHARE OF CROP AREA EQUIPPED FOR IRRIGATION (%)[b]	KILOGRAMS OF FERTILIZERS PER HECTARE OF CROPLAND	TRACTORS PER 1,000 FARM WORKERS	YEAR	EAST ASIA WITHOUT CHINA	CHINA
1960	0.0	0.0	0.0	21.1	8.2	0.2	**1990**	81,831	765,307
1980	40.9	61.7	27.5	31.0	102.3	1.9	**2000**	78,400	1,279,533
2000	80.5	89.6	89.1	32.1	189.3	2.3	**2010**	153,314	1,758,000
2015	—	—	—	34.6	258.1	11.1	**2017**	140,625	1,763,000

Sources: Original table for this publication. Cropland, irrigated area, fertilizer, and tractor use from FAO 2019; adoption of modern crop varieties from Evenson and Gollin 2003; pesticide data from FAOSTAT (Statistics Division, Food and Agriculture Organization of the United Nations), 2020 (http://www.fao.org/faostat/en/#data).
Note: East Asia data for pesticide totals use in Indonesia, Lao PDR, Malaysia, Myanmar, Thailand, Timor-Leste, and Vietnam. — = not available.
a. Although wheat has been grown in Southeast Asia, it is not at present a significant crop. Changing circumstances, for example, the need to diversify as a result of overproduction of rice, have led to increasing attention being directed to its possibilities.
b. Area equipped to provide water (via irrigation) to crops. It includes areas equipped for full or partial control irrigation, equipped lowland areas, and areas equipped for spate irrigation (FAO AQUASTAT, 2020, http://www.fao.org/aquastat/statistics/query/index.html).

TABLE 4.2 **Rice and maize yields, irrigated cultivated area, and fertilizer and pesticide use in developing East Asia**

	RICE YIELD (KG/HA)		MAIZE YIELD (KG/HA)		IRRIGATED AREA (% OF CULTIVATED AREA)	FERTILIZER USE (N-P-K) (KG/HA)		PESTICIDE USE (KG/HA)	
	1961	2018	1961	2018		2002	2017	1990	2017
Cambodia	1,092	3,570	1,979	4,896	9.1	6	21	—	—
China	2,079	7,028	1,185	6,104	51.5	318	390	5.87	13.07
Indonesia	1,762	5,191	927	5,326	16.0	67	114	0.08	0.03
Lao PDR	871	4,226	1,700	5,927	25.9	—	—	0.02	0
Malaysia	2,109	4,076	1,201	7,459	5.0	161	183	6.08	8.1
Myanmar	1,607	3,790	809	3,841	18.1	4	34	0.02	1.37
Philippines	1,230	3,971	628	3,094	17.3	75	88	—	—
Thailand	1,659	3,093	2,008	4,504	33.8	94	111	0.91	1.66
Vietnam	1,897	5,818	113	4,720	48.7	222	276	3.28	1.66

Sources: Original table for this publication. Pesticide data, fertilizer use, rice yield, and maize yield from FAOSTAT (FAOSTAT (Statistics Division, Food and Agriculture Organization of the United Nations, http://www.fao.org/faostat/en/#data); irrigated land area from FAO Irrigation Statistics.
Note: Irrigated land area: FAO Irrigation Statistics values range from 2005 to 2017: China (2013), Malaysia (2009), Thailand (2017), Indonesia (2005), Philippines (2017), Vietnam (2005), Lao PDR (2017), Cambodia (2006), Myanmar (2014), Mongolia (2017). — = not available; kg/ha = kilograms per hectare; N-P-K = nitrogen, phosphorus, and potassium.

Plantation crops and aquaculture have gained ground in developing East Asia in land area, value, and environmental and social costs. The region includes several significant plantation crop producers such as the world's largest exporters of rubber (Indonesia, Malaysia, Thailand, Vietnam), palm oil (Indonesia, Malaysia), coffee (Indonesia, Vietnam), and aquaculture and fisheries (China, Indonesia, Vietnam).[3] Significant area expansion, including deforestation and the growing of crops in more marginal and erosion-prone land, and production intensification (fertilizer, pesticide, water) have contributed to this increase in production and yields (table 4.3) and in trade and market share—as well as to the considerable environmental and social costs that accompanied these advances (box 4.1).

TABLE 4.3 **Coffee, palm oil, and rubber production**

COMMODITY AND COUNTRY	AREA HARVESTED (1,000 HA)		YIELD (KG/HA)		PRODUCTION (KILOTONS)		EXPORT VALUE (US$, MILLION)
	1961	2018	1961	2018	1961[a]	2018	2017
Coffee–Indonesia	182	1,242	567	582	103	722	1,176
Coffee–Malaysia[b]	7.9	2.3	339	3,767	2.7	8,7	2.6
Coffee–Vietnam	21	619	193	2,610	4.1	1,616	3,079
Palm oil–Indonesia	—	—	—	—	146	40,567	18,513
Palm oil–Malaysia	—	—	—	—	945	19,516	9,660
Palm oil–Thailand	—	—	—	—	0.6	2,777	217
Rubber–Indonesia	1,353	2,400	512	989	693	3,630	9.7
Rubber–Malaysia	1,300	1,104	607	708	790	782	78
Rubber–Thailand	400	3,230	465	1,469	186	4,744	1,526
Rubber–Vietnam	123	689	636	1,650	78	1,138	118
Aquaculture–China	—	—	—	—	29,750	66,135	—
Aquaculture–Indonesia	—	—	—	—	994	14,772	—
Aquaculture–Vietnam	—	—	—	—	514	4,153	—

Sources: Original table for this publication. Data on crops from FAOSTAT (Statistics Division, Food and Agriculture Organization of the United Nations), 2020 (http://www.fao.org/faostat/en/#data); data on aquaculture from Organisation for Economic Co-operation and Development (OECD) aquaculture production data.

Note: — = not available; ha = hectare; kg/ha = kilograms per hectare.

a. The year 1961 was chosen to match the onset of the Green Revolution. A production surge may have begun later, for example, coffee (green) production surged in Vietnam from the 1980s onward. Data for aquaculture is for 2000 and 2018 (OECD aquaculture production data).

b. Malaysia is not a major coffee producer; however, its coffee yields have increased significantly.

BOX 4.1

Environmental and societal challenges associated with plantation crops and aquaculture in developing East Asia

Large tea, coffee, palm oil, and other plantations tend to lose soil nutrients and accumulate agrochemicals where poor management practices are followed, resulting in loss of soil fertility and soil biodiversity. Cultivation on steep hills or mountainsides increases the potential for landslides and can markedly decrease water quality with up to fourfold increases in sediment loads in streams (Edwards et al. 2014).

Plantation crops and shrimp farming are also a leading cause of biodiversity loss, deforestation, and habitat loss. The Lower Mekong Basin region (parts of Cambodia, China, the Lao People's Democratic Republic, Thailand, and Vietnam) houses more than 20,000 plant species, 430 mammals, 1,200 birds, and abundant reptile and amphibian species and is considered the largest freshwater fishery in the world (ICEM 2013; Johnston et al. 2009). This abundant biodiversity often overlaps with areas that are suitable for agricultural production, ratcheting up risks of species and habitat loss. Between 29 percent and 42 percent of tropical deforestation around the world from 2000 to 2012 was related to commodity agriculture and timber export, with 32 percent of the total affected area in Asia (Lawson 2014).

Forest clearing for food or tree crop production is associated with loss of food security for local communities, loss of pollination services, loss of biodiversity, loss of carbon stores, increased pollution from fires for land clearing, and reduced flood retention capacity in riparian and coastal areas (Wertz-Kanounnikoff and Kongphan-Apirak 2008). Where expanding agriculture (especially shrimp farming) encroaches on mangrove systems, coastal and inland communities are increasingly vulnerable to soil erosion, salinization of inland water supplies, storm surges and tsunamis, and other climate-related risks (Barbier 2006; Scherr et al. 2015).

LIMITED TRADE-OFFS BETWEEN AGRICULTURAL INNOVATIONS THAT FOSTER ENVIRONMENTALLY SUSTAINABLE PRODUCTION AND PRODUCTIVITY

Intensive agriculture has generated environmental, social, and health costs. The Green Revolution made important environmental contributions by saving large areas of forest and woodland areas from conversion to agriculture (Stevenson et al. 2013).[4] However, it also generated environmental problems of its own (for example, nutrient pollution and pesticide residues), undermining the long-term sustainability of some intensive farming systems. Inappropriate management of modern inputs by farmers was the primary cause, and the problem was exacerbated by inadequate extension and training, ineffective regulation of water quality, and input pricing and subsidy policies that made modern inputs too cheap and encouraged excessive use of purchased inputs (Hazell 2009). Modifications in production systems, resource management, and spatial patterns of land use can reduce these trade-offs and generate synergies between environmental, economic, and social benefits (Buck et al. 2007; Milder et al. 2012; Scherr and McNeely 2008).

All developing East Asian countries have pursued sustainable cropping and livestock practices to some degree over the past 20–30 years. Both international agricultural research centers' and national agricultural research systems' breeding programs have since attempted to develop MVs that are less dependent on purchased inputs, and considerable effort has been devoted to research on farming systems, agronomic practices, integrated pest management (IPM), and other environment-friendly technologies. Details on a range of practices that are typically incorporated into environmentally sustainable production systems are featured in box D.3 in appendix D.

Most research and development and extension efforts for sustainable practices have targeted cereals. Because of its widespread cultivation and consumption, rice generates the greatest environmental impacts. Perhaps the most famous of the integrated sustainable approaches is IPM farmer field schools (FFS), which have been promoted in all East Asian countries (box 4.2). The other main integrated approaches that have gained ground in the region include the system of rice intensification, alternate wetting and drying, and practices promoting climate-smart agriculture (box 4.2). For other rice management approaches, see table D.1 and box D.5 in appendix D.

In crop farming, various field management practices (use of water, nutrients, pesticides) have demonstrated effectiveness in reducing reliance on inputs and negative environmental effects. In China, integrated soil-crop system management approaches have resulted in promising yield and profit results in maize, rice, and wheat production (box 4.3) (Cassou, Jaffee, and Ru 2017). In addition, countries such as Vietnam have pursued various approaches to improving input use efficiency (mostly pesticide and fertilizer) and use of biofertilizers and biopesticides (box 4.3). Precision application of inputs—according to crop needs—has also been tested and incrementally adopted in the region (box D.4 in appendix D). Such precision technologies may make a big difference in protecting the environment and rural communities, even in places with limited access to data and analytic tools, or where farmer sophistication is limited.[5]

Integrated livestock production practices have the potential to reduce both the environmental footprint and greenhouse gas (GHG) emissions associated with livestock. Despite livestock's importance as a livelihood and nutrition

BOX 4.2

Adoption of and constraints associated with selected sustainable rice production practices

Integrated pest management farmer field schools (IPM FFS) began in the 1980s and have been promoted in all East Asian countries. FFS are a common approach used to transfer specialist knowledge, promote skills, and empower farmers around the world. At least 10 million farmers in 90 countries have attended such schools. FFS are implemented by facilitators using participatory "discovery-based" learning based on adult education principles (Waddington et al. 2014). A meta-analysis verified that IPM FFS have been beneficial in improving intermediate outcomes relating to knowledge learned and adoption of beneficial practices, as well as final outcomes relating to agricultural production and farmers' incomes (Waddington et al. 2014). Their wider adoption, however, remains limited—FFS are a knowledge-intensive approach, which makes them challenging to convey to multiple small farmers, often with limited education (Waddington et al. 2014; World Bank 2005).

System of rice intensification (SRI) is an agro-ecological methodology for increasing the productivity of irrigated rice by changing the management of plants, soil, water, and nutrients. Over the past 30 years, SRI has evolved into a suite of principles, including reducing plant population, improving soil conditions and irrigation methods, and improving plant establishment methods, that are adaptable to local conditions. An estimated 4–5 million farmers in more than 50 countries around the world use SRI in whole or in part, with yield increases of 20–50 percent, reductions in required seed, and significant water savings (Uphoff 2012). SRI management leads to improved root structures, making crops more resilient to droughts, storms, and other climatic shocks (Abraham et al. 2014). Challenges to adoption include access to training and technical follow-up, labor needs, and access to tools and equipment (Styger and Uphoff 2016).

Governments have implemented SRI programs in each of the four countries producing two-thirds of the world's rice: China, India, Indonesia, and Vietnam (Uphoff 2012). In China, SRI is becoming the main rice cultivation system in much of the southern part of the country (Sichuan). These SRI production areas are using less water and fertilizer and are simultaneously boosting production beyond the national average of 6.6 tons per hectare to achieve 8–11 tons per hectare (IRIN 2012). In Vietnam, more than 1 million farmers were reportedly using SRI by 2011, representing 10 percent of all rice growers in the country, after initial studies demonstrated healthier rice fields and plants, reduced production costs, reduced water consumption, and increased profit (Dzung 2012).

Alternate wetting and drying is a water-saving irrigation technology that reduces water requirements by up to 40 percent and greenhouse gas emissions by 20–90 percent. Although alternate wetting and drying has resulted in savings in water use and reduction in greenhouse gases (Lampayan et al. 2015; Linquist, Snyder, and Anderson 2015), it has not been widely adopted because of its negative effects on yield (5 percent) and the complexity of the system (Carrijo, Lundy, and Linquist 2016).

source, its development has received less attention than crops. Production efficiency of livestock has increased, along with nutrient pollution (from manure) and GHG emissions[6] (FAO 2019). Improved livestock management practices can, however, reduce GHG emissions by about 30 percent (FAO 2019). Sustainable livestock production practices center on improved organizational strategies and technological innovations such as improved feeding, genetics, animal health, general husbandry, and information and communication technology use, which are driving up productivity and making resource use more efficient (FAO 2019). Regarding GHGs, four key approaches for reducing on-farm livestock emissions are currently available: (1) improving feed quality and digestibility, (2) precision farming of ruminant production,[7] (3) improving animal health and husbandry,

BOX 4.3

Integrated soil-crop management practices

Integrated soil-crop management practices can deliver higher yields and profits. In China, field experiments demonstrate that integrated soil-crop management practices, including reduction of nitrogen use by 4–14 percent in maize, rice, and wheat systems, can increase yields by 18–35 percent. Yields were increased from 7.2 metric tons per hectare to 8.5 metric tons per hectare without any increases in nitrogen fertilizer. Thus, a country with comparatively high rice yields can still increase productivity and reduce greenhouse gas emissions by managing nitrogen losses and using agronomic practices that control crop varieties, sowing dates, planting densities, and nutrient management practices (Chen et al. 2014).

Vietnam has implemented different input management programs with varying results. The "1 Must and 5 Reductions" (1M5R) program in Vietnam has shown that it could save Mekong Delta farmers an estimated 18–25 percent of their production costs per hectare without harming yields (Cassou, Jaffee, and Ru 2017). Developed by the International Rice Research Institute in collaboration with An Giang authorities, 1M5R calls for farmers to use certified seeds (the "1 Must") while reducing the use of four production inputs (seed, water, pesticides, and chemical fertilizers) and postharvest losses (the "5 Reductions"). Based on pilot studies through nine cropping seasons during 2012–14, Nguyen et al. (2015) estimate that 1M5R can potentially save farmers US$1.4 billion (32,000 billion Vietnamese dong) a year, assuming 4 million hectares of double-cropped rice. Similar opportunities for yields, incomes, and the environment have also been identified in relation to the use of fertilizer (and water) to grow coffee in Vietnam's Central Highlands region (Amarasinghe et al. 2015; TechnoServe 2013).

The "Three Reductions, Three Gains" (3R3G) project aimed to reduce production costs, improve farmers' health, and protect the environment in irrigated rice production in Vietnam through the reduced use of seeds, nitrogen fertilizer, and pesticides. It was developed by the International Rice Research Institute and introduced to farmers in southern Vietnam in early 2000 through traditional extension work and mass media. Farm survey data provide evidence of the adoption of 3R3G primarily in lowering seed rates (Huelgas, Templeton, and Castanar 2008).

A "small farmer, large field" model in Vietnam improves economies of scale. Small farmers often have limited means with which to invest in soil tests, power tools, and other technologies that enable precise application of chemicals. With limited landholdings, smallholders lack the opportunity to spread their fixed costs. The model has been promoted by province-level agricultural authorities to devise ways for small farmers to realize economies of scale without giving up their land rights. The model supports groups of 25–100 neighboring farms in managing the land as if it were a single farm. Farmers break down the walls between their plots, prepare the land together, manage water jointly, and plant the same varieties. Agriculture departments have encouraged farmers to form groups, facilitated contracts between groups and rice millers, and financed or provided land leveling, advisory, and other services (Cassou, Jaffee, and Ru 2017).

and (4) manure management of both ruminant and monogastric animals (table D.2 in appendix D). Improving overall energy efficiency is also a cost-effective option but generally works best in intensive and industrial-scale production systems, or when coupled with on-farm biogas production and energy generation (GRA and SAI 2013).

Biotechnology has shown significant potential for reducing agriculture's reliance on some of the most toxic pesticides. Genetically modified (GM) crops have been the fastest-adopted crop technology in the history of modern agriculture. Globally, the average impact of GM crop adoption was to reduce pesticide use by an average of 37 percent, increase crop yields by an average of 21 percent, and increase farm profits by an average of 69 percent.[8] The impacts in developing low- and middle-income countries were found to be larger than impacts in high-income countries (Klümper and Qaim 2014). The adoption of GM crops, for example, cotton and maize, that are pest tolerant or require less pesticide has also been fast and widespread in the East Asian countries that have endorsed them (table D.1 in appendix D). For example, the widespread adoption of *Bacillus thuringiensis* (Bt) cotton in China resulted in increased yields while reducing pesticide use and its negative health effects (box 4.4).

Organic agriculture is the most popular alternative farming system in the world. It is practiced in 162 countries and occupies about 1 percent of global cropland. The global sales of organic foods and beverages have grown from $63 billion in 2011 to nearly $166 billion in 2018 and are expected to reach about $320 billion by 2022. In 2011, most of these sales (90 percent) were concentrated in North America and Europe, with Asia,[9] Latin America, and Africa being primarily export producers (Willer, Lernoud, and Home 2013). Organic agriculture applies many sustainable production practices and delivers food with no pesticide residues (Smith-Spangler et al. 2012), greater energy efficiency and diversity, and enhanced soil carbon and quality (Alföldi et al. 2002; Mondelaers, Aertsens, and Van Huylenbroeck 2009). Yet it has received much criticism for inefficiency (Pickett 2013), lower yields (de Ponti, Rijk, and van Ittersum 2012), higher labor costs (Crowder and Reganold 2015), and reliance on more land to produce the same amount of produce (Trewawas 2001). Some recognize organic agriculture as being important for future food security, whereas others project it to become irrelevant (Crowder and Reganold 2015). A meta-analysis of organic agriculture concludes that, despite lower yields (but higher premiums), organic agriculture was more profitable than conventional agriculture (Crowder and Reganold 2015). The large-scale potential of organic agriculture in East Asia, where both agricultural land and labor are increasingly limited, must be weighed against productivity, efficiency, and consumer demand for healthy foods. Organic agriculture may have greater potential, for example, within controlled conditions of vertical farming and rapidly growing urban agriculture in general (chapter 5).

Irrigation has an important role in improving productivity, sustainability, and resilience of agriculture in increasingly fragile water and climatic conditions. Irrigation played a significant part in the adoption of MVs (Alaofè et al. 2016; Hazell 2009) as well as in the expansion of plantation crops in East Asia (table 4.3). Improvements in water management, including in irrigation services and training, have also been important for enhancing agricultural production and food security (Bizikova et al. 2017; Darko et al. 2016). Given weather and climatic risks, it is expected that one-half to two-thirds of future gains in global crop production will come from irrigated land (Kadiresan and Khanal 2018).

Technical and social innovations in water management and policy are required to improve water use efficiency (Rosegrant and Cai 2002; Yang, Zhang, and Zehnder 2003). In most parts of the world, variations of field flooding are still used. More efficient methods, such as central pivot sprinkler systems, have been in use in high-income countries since the 1950s. Drip irrigation, that is, slow application of water directly to crops, can minimize evaporation and reach water efficiency levels of 95 percent. Adding sensors can further reduce water use by as much as 50 percent and increase yields by 10 percent or more (Goldman Sachs 2016; Cornell University, INSEAD, and WIPO 2017). Farmers in northern China, where 75 percent of crop output is in irrigated land, increasingly resort to using water-saving plastic mulch or film as part of irrigation and cultivation methods (Ingman, Santelmann, and Tilt 2015), but have also greatly increased the use of groundwater (Wang et al. 2007; Zhang et al. 2008). Studies of the efficiency and impact of user-based, participatory management through water users associations have found that well set up and managed water users associations can promote water savings (15–20 percent) with no negative effects on farm incomes (Wang et al. 2005; Wang et al. 2006; Wang et al. 2010; Zhang et al. 2013). For details, see box D.6. in appendix D.

Climate-smart agriculture covers a broad range of innovations, services, and approaches that help farmers and the sector adapt to and mitigate climate change effects. Technologies such as solar-powered irrigation; drip irrigation; smarter crop-planting strategies; flood-, drought-, and disease-resistant varieties; and conservation agriculture can help farmers work within climate and environmental constraints and adapt to climate and environmental change. In Asia, climate-smart agriculture now also includes state-of-the-art simulation modeling technologies that show the impact of storm surges and sea-level rise, as well as field-focused technologies such as laser land leveling and solar-powered irrigation. In addition, countries have adopted climate-smart patterns of crop growth, optimizing planting dates based on improved intelligence about weather conditions (Dikitanan et al. 2017; Fuglie et al. 2020; Green 2018; Nguyen et al. 2017). Tables 4.4 and 4.5 summarize the status of climate-smart agriculture practice adoption for key commodities in Vietnam and the Philippines. Appendix D offers more details.

TABLE 4.4 **Climate-smart agriculture practices: Adoption rates for key crops and commodities in Vietnam**

CROP OR COMMODITY	PERCENT OF HARVESTED AREA	CLIMATE-SMART AGRICULTURE PRACTICE	ADOPTION RATE (%)
Rice	77	SRI	30–60
		Flood-resistant varieties	> 60
Maize	11	Drought-tolerant varieties	30–60
		Integrated pest management	< 30
Coffee	6	Water saving	30–60
		Intercropping with perennials	30–60
Shrimp	6	Shrimp-forest farming	30–60
		Shrimp-tilapia farming	< 30
Rubber	5	Agroforestry of rubber with legumes, upland rice, sesame	30–60
		Improved varieties tolerant to drought and frost	30–60
Cassava	5	Intercropping with cinnamon and leguminous species	30–60
		Mulching with leguminous species	< 30
Cashew	3	Drought-tolerant varieties	30–60
		Intercropping with coffee, pineapple, cocoa, perennial peanut, and others	30–60
Pork	n.a.	Biogas technology	< 30/30–60
		Use of local feed	30–60

Source: Adapted from Nguyen et al. 2017.
Note: n.a. = not applicable; SRI = system of rice intensification.

TABLE 4.5 **Climate-smart agriculture practices: Adoption rates for key crops and commodities in the Philippines**

CROP OR COMMODITY	PERCENT OF HARVESTED AREA	CLIMATE-SMART AGRICULTURE PRACTICE	ADOPTION RATE (%)
Rice	32	Water harvesting	< 30
		Site-specific nutrient management and pest management	
Maize	18	Site-specific nutrient management and pest management	< 30
		Early maturing and stress-tolerant varieties	
Integrated farming	27	Agroforestry systems (tree crops, rice, vegetables)	< 30
		Soil and water conservation techniques	
Livestock	n.a.	Alternative feeds	30–60
		Biogas and composting	< 30
Vegetables	5	Organic farming	< 30
		Drought-resistant varieties	30–60

Source: Adapted from Dikitanan et al. 2017.
Note: n.a. = not applicable.

CHALLENGES TO SMALLHOLDERS' ADOPTION OF INNOVATIONS FOSTERING SUSTAINABILITY

The scaling up of many promising innovations among millions of scattered smallholders has met with obstacles in the region. When a farmer chooses to adopt a new variety to replace an older variety, it reflects the farmer's judgment that the new variety offers a net benefit or advantage over the old one. Many farmers found it profitable to use MVs with high responsiveness to chemical fertilizers. Although farmers were generally quick to adopt MVs (Evenson and Gollin 2003) and GM crops (Qiao 2015; Zhang et al. 2018), the adoption of more complex and knowledge- or management-intensive practices (for example, IPM, system of rice intensification, organic agriculture, water users associations) has at times been difficult for smallholders (Cassou, Jaffee, and Ru 2017; Crowder and Reganold 2015; Uphoff 2012; Waddington et al. 2014). Similarly, practices that require a large up-front investment (for example, irrigation, biogas) have experienced slower uptake among smallholders who struggle to access finance or who do not have secure land rights (Cassou, Jaffee, and Ru 2017; Zhang et al. 2013).

Integrated production practices can generate a triple win for farmers and society, but farmers are not keen to adopt such approaches for several reasons. Box 4.4 discusses factors affecting innovation adoption in the Green Revolution context and compares it to the present day, that is, adoption of various sustainable crop management practices, including pest management by Bt.

Increasing the capabilities of farmers has, however, paid off—and returns have been greater in the presence of adaptable technologies and sectoral transformation. Post–Green Revolution technologies often aim to improve resource-use efficiency and product quality, and place significant demands on financial and risk management, marketing, and negotiating skills (for example, for input and output prices). Participation in commodity-specific out-grower schemes also requires organizational skills and the ability to comply with

BOX 4.4

Adoption of innovations in the Green Revolution context versus the triple win context: The nature of innovation, capacity, and the enabling environment matters

Improved cereal varieties, fertilizers, irrigation (in East Asia, more than 20 percent of fields were under irrigation at the onset of the Green Revolution), and modern pest control methods supported by extension and markets lay at the heart of the Green Revolution (Evenson and Gollin 2003; Hazell 2009). A scale-neutral technology package was conducive to widespread farmer adoption, but the Green Revolution was much more than a technology fix. This early adoption of modern crop varieties and synthetic fertilizers did not generally require more than basic literacy skills; however, it depended on a supporting extension system (farmer education, access to inputs) and markets (product absorption, price). Meeting these requirements typically required proactive efforts by governments in the form of land reforms, small-farm development programs, and input and credit subsidies (Hazell 2009). In China and Vietnam, for example, the transformation of the agriculture sector was initiated with a significant policy focus on land tenure reform, opening of markets, infrastructure investments (rural roads, irrigation), along with investments in innovation—specifically agricultural technology and advisory services (Huang and Rozelle 2018; World Bank 2020). The widespread use of improved varieties also stimulated innovation in Asia's agricultural machinery industry (Huang, Rozelle, and Hu 2007).

Post–Green Revolution innovations focus on productivity, sustainability, and food safety (the triple win) but are more knowledge-intensive and place a greater emphasis on management and new skills. The uptake and spread of more sustainable technologies and farming practices in Asia's Green Revolution areas have remained inadequate (Waddington et al. 2014). Innovations geared toward sustainability (such as integrated pest management, the system of rice intensification, Vietnam's eco-practices in its rice program) are essentially about gradual adjustments to agronomic practices that rely on an understanding of the cropping system. These new innovations are increasingly management- and skill-intensive (Gollin, Morris, and Byerlee 2005). Indeed, in Vietnam's Mekong Delta, farmers have often resisted the adoption of eco-friendly practices in rice. Adoption of eco-practices was, on the one hand, positively correlated with improved access to information and knowledge (as evidenced by membership in farmer organizations, perceived ease of use, difference in selling price, farmer experience, perception of biodiversity losses). On the other hand, farmers' risk perceptions and the number of paddy plots negatively affected its adoption (Vo et al. 2018). There are several other possible reasons for inadequate adoption, including perverse incentives caused by input subsidies, labor constraints, insecure property rights, difficulty of organizing for economies of scale, and externality problems (Hazell 2009; Scherr et al. 2015).

Bacillus thuringiensis (Bt) cotton is a new technology integrated with existing cotton production practices. China is the largest cotton producer in the world. Historically, about a third of pesticides in China are used on cotton, and many of these are classed as extremely hazardous by the World Health Organization, contributing to 400–500 farmer deaths annually from pesticide poisoning. To address this threat, government research systems developed genetically modified Chinese Bt insect-resistant cotton and introduced it to Chinese cotton farmers in 1997. It reached full adoption by 2012 (Zhang et al. 2018). Both cotton bollworm infestations and the use of insecticide sprays to control the pest declined dramatically between 1997 and 2015. A resurgence in beneficial insects and arthropod predators was observed. Since 2008 both spraying and pest infestations have shown a slightly decreasing trend, suggesting that cotton production in China is steadily becoming less environmentally destructive (Qiao 2015; Zhang et al. 2018). Although the health benefits, reduced pesticide use (55 percent), improved yield, and environmental improvements are undisputed (Huang et al. 2002), adopting genetically modified Bt crops has not been a Green Revolution–type silver bullet in reducing pesticide use. The practice of applying excessive amounts of highly toxic pesticides has continued even after the adoption of Bt cotton (Zhang et al. 2018). Behavioral factors such as risk aversion and lack of knowledge by Chinese cotton farmers can be important factors driving pesticide use, and others have suggested that

continued

Box 4.4, *continued*

Bt cotton seed quality and secondary pests are also at play (Lynas 2018; Qiao 2015; Zhang et al. 2018). Some evidence also suggests that small farms (in Indonesia and China) may be more prone to wasting agrochemicals than larger or commercial farms (Ju et al. 2016; Osorio et al. 2011). In conclusion, although Bt cotton has generated immense benefits, partly overcoming the challenges with more management- and knowledge-intensive innovations (for example, integrated pest management), farmer adoption is still hampered for a variety of reasons, including limited information and knowledge.

commodity-specific quality and safety standards. Acquiring the numeracy, information and communication technology competency, and other skills necessary in this environment will require continually upgrading the formal education levels of the farm workforce (Fuglie et al. 2020). One consistent finding from technology adoption studies is that more-educated farmers adopt new technologies (of any kind) early and get more profit out of them (Foster and Rosenzweig 2010). Given that education facilitates the acquisition and processing of new information, the impacts of education on farm productivity and income are higher in environments undergoing technological and structural transformation (Lockheed, Jamison, and Lau 1980). Raising the human capital of farmers (basic education, vocational training, extension) allows them to better evaluate technological opportunity and manage technology-related investments. Unsurprisingly, the returns to education increase when there are greater opportunities for new technological adoption (Fuglie et al. 2020).

Governments have also used various incentives to address market failures that have prevented the adoption of more sustainable agriculture practices. Such market failures include externalities, transaction costs, and information asymmetries (box 4.5). Approaches to help provide incentives for more sustainable agricultural practices can be divided into three groups: the creation and

BOX 4.5

Incentives or disincentives tied to improved farming practices and agro-industry services

China, the European Union, India, the Republic of Korea, Vietnam, and the United States have used different approaches to foster the greening of agriculture. The list of options below is drawn from these experiences.

- Fines or loss of benefits for noncompliance with mandates
- Preferential credit or grants for straw residue management or manure injection machinery
- Subsidies for formula fertilizer, fertilizer deep-placement products, or soil testing kits
- Payments for adopting practices that reduce farm runoff
- Public procurement requirement that food purchases meet given certification standards
- Fast-track licensing for operations meeting high environmental management standards
- Loans to enterprises offering input application and soil testing services, as well as improved drugs, inputs, and gear
- Grants for demonstration farms and farmer-led movements modeling and supporting best practices and to enterprises
- Increasing access to and the appeal of plant-based or low-footprint foods

Source: Cassou, Jaffee, and Ru 2017.

ramping up of agri-environmental subsidy programs (as in China, the Republic of Korea, and the United States), leveraging existing subsidies to advance environmental objectives (as in China, the European Union, India, Vietnam, and the United States), and reform of subsidies (for example, on fertilizer) that have an indirect effect on the environment (as in China and Korea).

NOTES

1. In the late Green Revolution period (1981–2000), production gains were more dependent on MVs than they had been in the early period when productivity growth was also accelerated by expanding irrigation, fertilizer, and pesticide use, accounting for almost 50 percent of yield growth and 40 percent of production growth for all developing countries (Evenson and Gollin 2003).
2. Mechanization has contributed greatly to productivity growth in agriculture. The economies with the highest number of agricultural machines are China, India, Japan, Poland, and the United States, with China and India, respectively, accounting for 25 percent and 14 percent of world agricultural machinery in use. Brazil, France, Italy, Thailand, and Turkey also stand out in the use of machinery in agriculture (Cornell University, INSEAD, and WIPO 2017).
3. Agribusiness has expressed interest in Myanmar, which may become a producer of rubber and palm oil in the future (World Bank 2017).
4. The global crop area in 2004 would have been 17.9–26.7 million hectares larger without crop germplasm improvement since 1965. Of these, 12.0–17.7 million hectares would have been in developing countries, resulting in 2 million hectares of additional deforestation (Stevenson et al. 2013).
5. The continued dominance of small farm size has challenged farmers' and regulators' capacity to moderate pollution. In Vietnam, the average rice farmer cultivates about half a hectare, and 90 percent of coffee growers have less than two hectares (Cassou, Jaffee, and Ru 2017).
6. Direct emissions from livestock and feed production constitute some 80 percent of total agriculture GHGs—about 65 percent of this 80 percent is related to cattle (GRA and SAI 2013).
7. Precision livestock farming provides for the individual animal's needs in bigger herds, integrating health, genetics, feed, social behavior, and resource use and availability, which can be supported by sensor technology integrated into monitoring systems. Precision application of fertilizer and irrigation, aided by remote sensing of soil moisture and pasture growth and quality, can improve resource use efficiency. Precision livestock farming thus builds on and extends the individual approaches of optimizing feed quality and digestibility and animal health and husbandry (GRA and SAI 2013).
8. These figures are according to a meta-analysis of 147 studies worldwide on the impact of GM crops (Klümper and Qaim 2014).
9. Organic farmland: Vietnam 2.19 percent, Philippines 1.76 percent, China 0.61 percent, Cambodia 0.51 percent, Indonesia 0.44 percent, Thailand 0.43 percent, Lao PDR 0.32 percent, Malaysia 0.12 percent, and Myanmar 0.10 percent (FiBL Statistics 2018 data on organic area worldwide, https://statistics.fibl.org/world/area-world.html?tx_statisticdata_pi1%5Bcontroller%5D=Element2Item&cHash=f367262839ab9ca2e7ac1f333fbb1ca2).

REFERENCES

Abraham, B., O. O. AdeOluwa, H. Araya, T. Berhe, Y. Bhatt, S. Edwards, B. Gujja, et al. 2014. *The System of Crop Intensification: Agroecological Innovations for Improving Agricultural Production, Food Security, and Resilience to Climate Change.* Ithaca: SRI International Network and Resources Center and Technical Centre for Agricultural and Rural Cooperation.

Alaofè, H., J. Burney, R. Naylor, and D. Taren. 2016. "Solar-Powered Drip Irrigation Impacts on Crops Production Diversity and Dietary Diversity in Northern Benin." *Food and Nutrition Bulletin* 37 (2): 164–75.

Alföldi, Thomas, Andreas Fliessbach, Uwe Geier, Lucas Kilcher, Urs Niggli, Lukas Pfiffner, Matthias Stolze, and Helga Willer. 2002. "Organic Agriculture and the Environment." In *Organic Agriculture, Environment, and Food Security*, edited by N. E.-H. Scialabba and C. Hattam. Rome: Food and Agriculture Organization of the United Nations. www.fao.org/docrep/005/y4137e/y4137e00.htm.

Amarasinghe, U. A., C. T. Hoanh, D. D'haeze, and T. Q. Hung. 2015. "Toward Sustainable Coffee Production in Vietnam: More Coffee with Less Water." *Agricultural Systems* 136: 96–105. http://dx.doi.org/10.1016/j.agsy.2015.02.008.

Barbier, E. B. 2006. "Mangrove Dependency and the Livelihoods of Coastal Communities in Thailand." In *Environment and Livelihoods in Coastal Zones*, edited by C. T. Hoanh, T. P. Tuong, J. W. Gowing, and B. Hardy. Laramie, WY: CABA International Press.

Bizikova, L., L. S. Jungcurt, K. McDougal, and C. Smaller. 2017. "Effective Public Investments to Improve Food Security." International Institute for Sustainable Development, Winnipeg, Manitoba, Canada. https://www.iisd.org/sites/default/files/publications/effective-public-investments-improve-food-security.pdf.

Buck, L. E., T. A. Gavin, N. T. Uphoff, and D. R. Lee. 2007. "Scientific Assessment of Ecoagriculture Systems." In *Farming with Nature: The Science and Practice of Ecoagriculture*, edited by S. J. Scherr and J. A. McNeely. Washington, DC: Island Press.

Carrijo, R. D., M. Lundy, and B. A. Linquist. 2016. "Rice Yields and Water Use under Alternate Wetting and Drying Irrigation: A Meta-Analysis." *Field Crops Research* 203 (1): 173–80. https://doi.org/10.1016/j.fcr.2016.12.002.

Cassou, E., S. M. Jaffee, and J. Ru. 2017. *The Challenge of Agricultural Pollution: Evidence from China, Vietnam, and the Philippines*. Directions in Development. Washington, DC: World Bank.

Chen, X., Z. Cui, M. Fan, P. Vitousek, M. Zhao, W. Ma, Z. Wang, et al. 2014. "Producing More Grain with Lower Environmental Costs." *Nature* 514 (7523): 486–89. doi:10.1038/nature13609.

Cornell University, INSEAD, and WIPO (World Intellectual Property Organization). 2017. *The Global Innovation Index 2017: Innovation Feeding the World*, 10th ed. Ithaca: Cornell University; Fontainebleau: INSEAD; and Geneva: WIPO. https://www.insead.edu/sites/default/files/assets/dept/globalindices/docs/GII-2017-report.pdf.

Crowder, D. W., and J. P. Reganold. 2015. "Financial Competitiveness of Organic Agriculture on a Global Scale." *Proceedings of the National Academy of Sciences of the United States of America* 112 (24): 7611–16. doi:10.1073/pnas.1423674112.

Darko, R. O., S. Yuan, L. Hong, J. Liu, and H. Yan. 2016. "Irrigation, a Productive Tool for Food Security—A Review." *Acta Agriculturae Scandinavica. Section B. Soil and Plant Science* 66 (3): 191–206.

de Ponti, T., B. Rijk, and M. K. van Ittersum. 2012. "The Crop Yield Gap between Organic and Conventional Agriculture." *Agricultural Systems* 108 (April): 1–9.

Dikitanan, R., G. Grosjean, A. Nowak, and J. Leyte. 2017. "Climate-Resilient Agriculture in Philippines." CSA Country Profiles for Asia Series. International Center for Tropical Agriculture (CIAT); Department of Agriculture—Adaptation and Mitigation Initiatives in Agriculture, Government of the Philippines, Manila.

Dzung, N. 2012. "Simple and Effective SRI and Agriculture Innovation." SRI International Network and Resources Center, Ithaca.

Edwards, D. P., J. A. Tobias, D. Sheil, E. Meijaard, and W. Laurance. 2014. "Maintaining Ecosystem Function and Services in Logged Tropical Forests." *Trends in Ecology & Evolution* 29 (9): 511–20.

Evenson, R. E., and D. Gollin, eds. 2003. *Crop Variety Improvement and Its Effect on Productivity: The Impact of International Agricultural Research*. Wallingford, UK: CABI.

FAO (Food and Agriculture Organization of the United Nations). 2019. *Five Practical Actions towards Low-Carbon Livestock*. Rome: FAO.

Foster, A., and M. R. Rosenzweig. 2010. "Microeconomics of Technology Adoption." *Annual Review of Economics* 2 (1): 395–424.

Fuglie, K., M. Gautam, A. Goyal, and W. F. Maloney. 2020. *Harvesting Prosperity: Technology and Productivity Growth in Agriculture*. Washington, DC: World Bank.

Goldman Sachs. 2016. "Precision Farming: Cheating Malthus with Digital Agriculture." Equity Research, Profiles in Innovation, July 13. http://docdrop.org/static/drop-pdf/GSR_agriculture-N1sH6.pdf.

Gollin, D. 2010. "Agricultural Productivity and Economic Growth." In *Handbook of Agricultural Economics*, Vol. 4, edited by R. E. Evenson and P. Pingali, 3825–66. Amsterdam: Elsevier.

Gollin, D., M. Morris, and D. Byerlee. 2005. "Technology Adoption in Intensive Post-Green Revolution Systems." *American Journal of Agricultural Economics* 87 (5): 1310–16.

GRA (Global Research Alliance on Greenhouse Gases) and SAI (Sustainable Agriculture Initiative). 2013. "Reducing Greenhouse Gas Emissions from Livestock: Best Practice and Emerging Options." GRA and SAI. https://saiplatform.org/uploads/Modules/Library/lrg-sai-livestock-mitigation_web2.pdf.

Green, Adam. 2018. "Food 4.0: The Future of Food Innovation in Asia." Economist Intelligence Unit Limited.

Hazell, P. B. R. 2009. "The Asian Green Revolution." Discussion Paper 00911, International Food Policy Research Institute, Washington, DC.

Huang, J., and S. Rozelle. 2018. "China's 40 Years of Agricultural Development and Reform." Australian National University, Canberra, Australia.

Huang, J., S. Rozelle, C. E. Pray, and Q. Wang. 2002. "Plant Biotechnology in the Developing World: The Case of China." *Science* 295 (25): 674–77.

Huang, Q., S. Rozelle, and D. Hu. 2007. "Pump-Set Clusters in China: Explaining the Organization of the Industry That Revolutionized Asian Agriculture." *Asia-Pacific Development Journal* 14 (December): 75–105.

Huelgas, Z. M., D. Templeton, and P. Castanar. 2008. "Three Reductions, Three Gains (3R3G) Technology in South Vietnam: Searching for Evidence of Economic Impact." Contributed paper at the 52nd Annual Conference of the Australian Agricultural Resource Economics Society, Canberra, Australia, February 5–8.

ICEM (International Centre for Environmental Management). 2013. *USAID Mekong ARCC Climate Change Impact and Adaptation: Main Report*. Bangkok: USAID Asia Regional Environment Office.

Ingman, M., M. V. Santelmann, and B. Tilt. 2015. "Agricultural Water Conservation in China: Plastic Mulch and Traditional Irrigation." *Ecosystem Health and Sustainability* 1 (4): 1–11. doi:10.1890/EHS14-0018.1.

IRIN. 2012. "Analysis: Why Rice Intensification Matters in Asia." IRIN Middle East English Service, April 24. http://www.irinnews.org/report/95342/analysis-why-rice-intensification-matters-in-asia.

Johnston, R., C. T. Hoanh, G. Lacombe, A. Noble. V. Smakhtin, D. Suhardiman, S. P. Kam, and P. S. Choo. 2009. "Scoping Study on Natural Resources and Climate Change in Southeast Asia with a Focus on Agriculture." International Water Management Institute, South East and Southeast Asia Office, Vientiane.

Ju, X., B. Gu, Y. Wu, and J. N. Galloway. 2016. "Reducing China's Fertilizer Use by Increasing Farm Size." *Global Environmental Change* 41: 26–32.

Kadiresan, K., and P. R. Khanal. 2018. "Rethinking Irrigation for Global Food Security." *Irrigation and Drainage* 67 (1): 8–11.

Klümper, W., and M. Qaim. 2014. "A Meta-Analysis of the Impacts of Genetically Modified Crops." *PLoS ONE* 9 (11): e111629. https://doi.org/10.1371/journal.pone.0111629.

Lampayan, R. M., R. M. Rejesus, G. R. Singleton, and B. A. M. Bouman. 2015. "Adoption and Economics of Alternate Wetting and Drying Water Management for Irrigated Lowland Rice." *Field Crops Research* 170 (January): 95–108.

Lawson, S. 2014. *Consumer Goods and Deforestation: An Analysis of the Extent and Nature of Illegality in Forest Conversion for Agriculture and Timber Plantations*. Washington, DC: Forest Trends.

Linquist, R., R. Snyder, and F. Anderson. 2015. "Water Balances and Evapotranspiration in Water- and Dry-Seeded Rice Systems." *Irrigation Science* 33 (2015): 375–85.

Lockheed, M., D. Jamison, and L. J. Lau. 1980. "Farmer Education and Farm Efficiency: A Survey." *Economic Development and Cultural Change* 29 (1): 37–76.

Lynas, M. 2018. "GMO Cotton Prompts Dramatic Drop in China's Pesticide Use." Cornell Alliance for Science (blog), September 20, 2018. https://allianceforscience.cornell.edu/blog/2018/09/gmo-cotton-prompts-dramatic-drop-chinas-pesticide-use/.

Milder, J. C., K. Garbach, F. A. J. DeClerk, L. Driscoll, and M. Montenegro. 2012. "An Assessment of the Multi-Functionality of Agroecological Intensification." Report prepared for the Bill and Melinda Gates Foundation, Ecoagriculture Partners, Washington, DC.

Mondelaers, K., J. Aertsens, and G. Van Huylenbroeck. 2009. "A Meta-Analysis of the Differences in Environmental Impacts between Organic and Conventional Farming." *British Food Journal* 111 (10): 1098–119.

Nguyen, T. H., T. H. Tran, M. D. Chau, and V. S. Nguyen. 2015. "Monitoring and Evaluating of the Application of Low Carbon Rice Production Technologies 1M5Rs and Its Impacts in Kien Giang and Giang Provinces. Workshop Proceedings." Vietnam Low Carbon Rice Project (VLCRP), Project-End Dissemination and Regional Policy Dialogue Workshop, Kien Giang, April 15.

Nguyen, T. T. N., F. Roehrig, G. Grosjean, D. N. Tran, and T. M. Vu. 2017. "Climate Smart Agriculture in Vietnam." CSA Country Profiles for Asia Series, International Center for Tropical Agriculture (CIAT), Food and Agriculture Organization, Hanoi.

Osorio, C. G., D. E. Abriningrum, E. B. Armas, and M. Firdaus. 2011. "Who Is Benefiting from Fertilizer Subsidies in Indonesia?" Policy Research Working Paper 5758, World Bank, Washington, DC. http://documents.worldbank.org/curated/en/483961468039869443/pdf/WPS5758.pdf.

Pickett, J. A. 2013. "Food Security: Intensification of Agriculture Is Essential, for Which Current Tools Must Be Defended and New Sustainable Technologies Invented." *Food and Energy Security* 2 (3): 167–17.

Qiao, F. 2015. "Fifteen Years of Bt Cotton in China: The Economic Impact and Its Dynamics." *World Development* 70 (C): 177–85.

Rosegrant, M., and X. Cai. 2002. "Water Constraints and Environmental Impacts of Agricultural Growth." *American Journal of Agricultural Economics* 84 (3): 832–38.

Scherr, Sara J., Kedar Mankad, Steven Jaffee, and Christine Negra, with Tanja Havemann, Janjarang Kijtikhun, Uray Endang Kusumjaya, Samiksha Nair, and Naomi Rosenthal. 2015. *Steps toward Green: Policy Responses to the Environmental Footprint of Commodity Agriculture in East and Southeast Asia.* Washington, DC: EcoAgriculture Partners and the World Bank.

Scherr, S. J., and J. A. McNeely. 2008. "Biodiversity Conservation and Agricultural Sustainability: Towards a New Paradigm of 'Ecoagriculture' Landscapes." *Philosophical Transaction of the Royal Society B* 363: 477–94.

Smith-Spangler, C., M. L. Brandeau, G. E. Hunter, J. C. Bavinger, M. Pearson, P. J. Eschbach, V. Sundaram, et al. 2012. "Are Organic Foods Safer or Healthier Than Conventional Alternatives? A Systematic Review." *Annals of Internal Medicine* 157 (5): 348–66.

Stevenson, J. R., N. Villoria, D. Byerlee, T. Kelley, and M. Maredia. 2013. "Green Revolution Research Saved an Estimated 18 to 27 Million Hectares from Being Brought into Agricultural Production." *Proceedings of the National Academy of Sciences* 110 (21): 8363–68. doi:10.1073/pnas.1208065110.

Styger, E., and N. Uphoff. 2016. "The System of Rice Intensification (SRI): Revisiting Agronomy for a Changing Climate." Practice Brief, Climate-Smart agriculture, Global Alliance for Climate Smart Agriculture. https://cgspace.cgiar.org/bitstream/handle/10568/77040/GACSA%20Practice%20Brief-%20SRI%20FINAL.pdf.

TechnoServe. 2013. "A Business Case for Sustainable Coffee Production—Vietnam: An Industry Study by TechnoServe for the Sustainable Coffee Program." Report prepared for IDH, the Sustainable Trade Initiative, Hanoi.

Timmer, C. P. 2002. "Agriculture and Economic Development." In *Handbook of Agricultural Economics*, Vol. 2A, edited by B. Gardner and G. Rausser, 1487–546. Amsterdam: North-Holland.

Trewawas, A. 2001. "Urban Myths of Organic Farming." *Nature* 410: 409–10.

Uphoff, N. 2012. "Supporting Food Security in the 21st Century through Resource-Conserving Increases in Agricultural Production." *Agriculture and Food Security* 1 (18): 1–12.

Vo, Hong Tu, Nguyen Duy Can, Yoshifumi Takahashi, Steven W. Kopp, and Mitsuyasu Yabe. 2018. "Modelling the Factors Affecting the Adoption of Eco-Friendly Rice Production in the Vietnamese Mekong Delta." *Cogent Food & Agriculture* 4 (1): 1432538.

Waddington, H., B. Snilstveit, J. Hombrados, M. Vojtkova, D. Phillips, P. Davies, and H. White. 2014. "Farmer Field Schools for Improving Farming Practices and Farmer Outcomes: A Systematic Review." *Campbell Systematic Reviews* 10 (1): i–335. doi:10.4073/csr.2014.6.

Wang, J., J. Huang, S. Rozelle, Q. Huang, and A. Blanke. 2007. "Agriculture and Groundwater Development in Northern China: Trends, Institutional Responses, and Policy Options." *Water Policy* 9 (S1): 61–74.

Wang, J., J. Huang, L. Zhang, Q. Huang, and S. Rozelle. 2010. "Water Governance and Water Use Efficiency: The Five Principles of WUA Management and Performance in China." *Journal of the American Water Resources Association* 46 (4): 665–85.

Wang, J., Z. Xu, J. Huang, and S. Rozelle. 2005. "Incentives in Water Management Reform: Assessing the Effect on Water Use, Production, and Poverty in the Yellow River Basin." *Environment and Development Economics* 10 (6): 769–99.

Wang, J., Z. Xu, J. Huang, and S. Rozelle. 2006. "Incentives to Managers or Participation of Farmers in China's Irrigation Systems: Which Matters Most for Water Savings, Farmer Income, and Poverty?" *Agricultural Economics* 34 (3): 315–30.

Wertz-Kanounnikoff, S., and M. Kongphan-Apirak. 2008. "Reducing Forest Emissions in Southeast Asia: A Review of Drivers of Land Use Change and How Payments for Environmental Services (PES) Schemes Can Affect Them." Working Paper 41, Center for International Forestry Research, Bogor, Indonesia.

Willer, H., J. Lernoud, and R. Home. 2013. "The World of Organic Agriculture 2013: Summary." In *The World of Organic Agriculture: Statistics and Emerging Trends 2013*, edited by H. Willer, J. Lernoud, and L. Kilcher, 26–33. Bonn: FiBL-IFOAM.

World Bank. 2005. "Sustainable Pest Management: Achievements and Challenges." Agriculture and Rural Development Department, World Bank, Washington, DC.

World Bank. 2017. *Myanmar's Public Expenditure Review: Fiscal Space for Economic Growth.* Yangon: World Bank Myanmar.

World Bank. 2020. "The Future of Agriculture & Food Sector. A Scenario Analysis." Unpublished manuscript, World Bank, Washington, DC.

Yang, H., X. Zhang, and A. J. B. Zehnder. 2003. "Water Scarcity, Pricing Mechanism and Institutional Reform in Northern China Irrigated Agriculture." *Agricultural Water Management* 61 (2): 143–61.

Zhang, L., J. Wang, J. Huang, and S. Rozelle. 2008. "Development of Groundwater Markets in China: A Glimpse into Progress to Date." *World Development* 36 (4): 706–26.

Zhang, L., N. Heerink, L. Dries, and X. Shi. 2013. "Water Users Associations and Irrigation Water Productivity in Northern China." *Ecological Economics* 95 (November): 128–36. https://doi.org/10.1016/j.ecolecon.2013.08.014.

Zhang, W., Y. Lu, W. van der Werf, J. Huang, F. Wu, K. Zhou, X. Deng, Y. Jiang, K. Wu, and M. W. Rosegrant. 2018. "Multidecadal, County-Level Analysis of the Effects of Land Use, Bt Cotton, and Weather on Cotton Pests in China." *PNAS* 115 (33): E7700–E7709. https://doi.org/10.1073/pnas.1721436115.

5 Transformative Innovations for Resilience and Sustainable and Safe Growth

INTRODUCTION

Transformative innovations have the potential to redirect the development of the agri-food system. As discussed in chapter 2, the agri-food system in developing East Asia has become increasingly vulnerable to a variety of risks and external shocks and is currently not able to sustainably meet its growing demand for nutritious and safe food. While learning from the past, it is crucial to look to the future of the agri-food system. Several new transformative innovations across the entire agri-food chain—many geared toward greener growth and food safety—may have the potential to address both supply- and demand-side challenges of the agri-food system. However, they are yet to reach most producers and entrepreneurs.[1] These innovations can be divided into three main groups: innovations fostering (1) agricultural production (drawing on digital technology, biotechnology, or off-farm and alternative sources); (2) the functioning of the food chain; and (3) food consumption and nutrition (presented in table 1.4). These groups, however, overlap; for instance, lab-grown meat (meat produced by in vitro cell culture of animal cells) addresses demand for environmentally sustainable production as well as nutrition. Food quality is addressed in all three groups, for instance, through biotechnology, e-commerce, and reformulation. COVID-19 (coronavirus) impacts on the food chain have also brought about a greater sense of urgency with regard to the adoption of some transformative innovations (discussed in chapter 2).[2]

THE TRANSFORMATIVE ROLE OF TECHNOLOGY IN AGRICULTURAL PRODUCTIVITY

Agricultural productivity drawing on digital technology

The agriculture sector has experienced consecutive revolutions brought about by technology. The innovations of the Green Revolution greatly improved agricultural productivity. Subsequently, biotechnology and the growth of information and communication technology (ICT) have provided opportunities to overcome some of the challenges facing agriculture and more recently the overall agri-food system. The transformative impact of ICT on agriculture is widely recognized, with the public and private sectors rapidly adopting digital solutions to address agriculture challenges (figure 5.1). These technologies range from the traditional tools of the mobile phone, television, radio, and the Internet to recent applications using the Internet of things, big data analytics and information systems, drones and remote sensing (using geographic information systems [GIS]), to blockchain (FAO 2019a; World Bank 2017a).

FIGURE 5.1

Role of information and communication technologies in agriculture and examples of digital technology applications

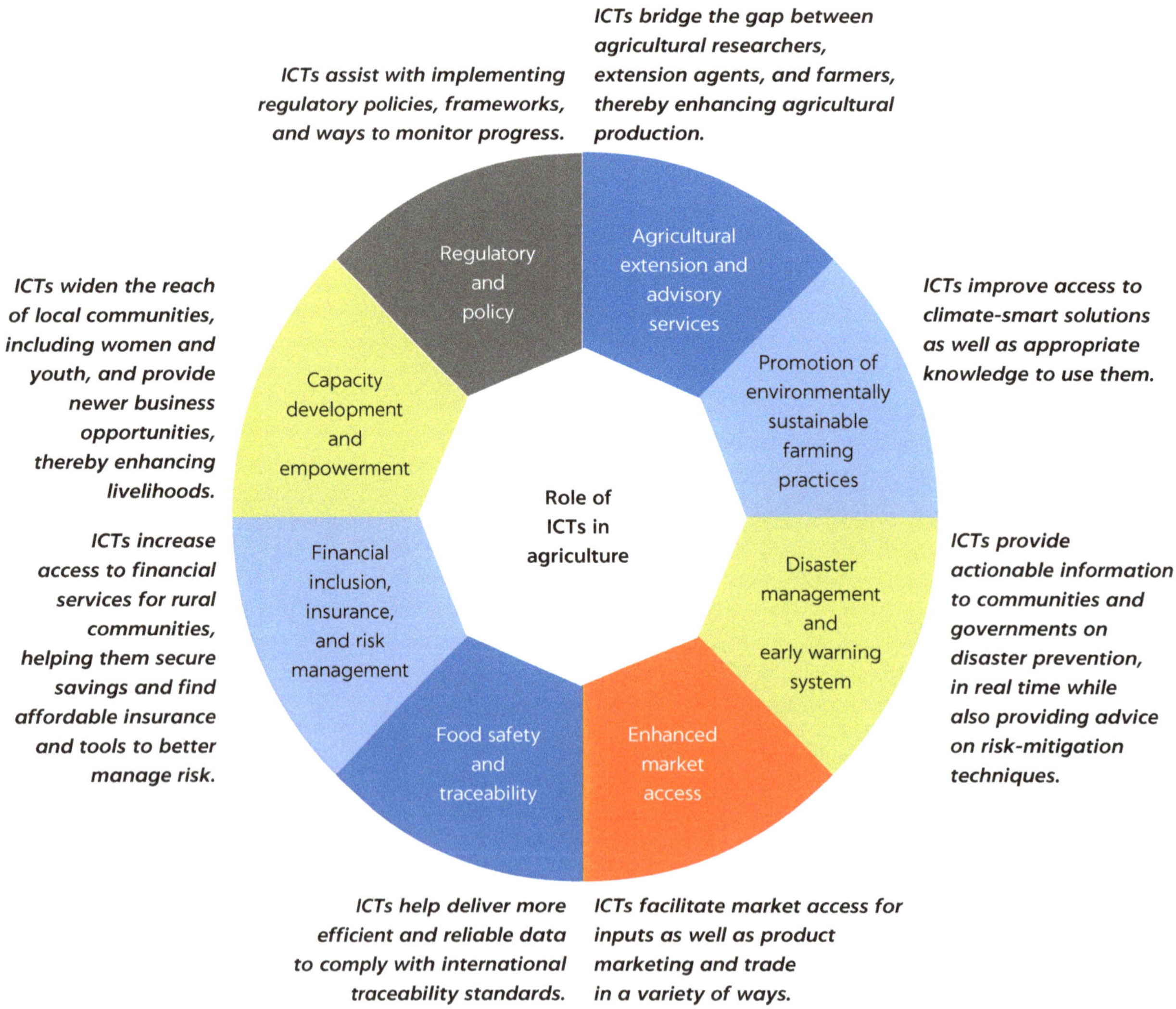

Internet of things: Checking soil health, introducing the traceability of products.
Big data analytics: Customized weather and agriculture advisory services, e-agriculture marketplace information, disaster alerts.
Blockchain: Smart contracts, improved supply chain monitoring, food safety, insurance.
Drone and GIS-based applications: Land use mapping, crop monitoring, productivity estimation, weather advisory services.
Artificial intelligence: Plant disease detection, weather prediction, climate change analytics.

Source: FAO 2019a.
Note: GIS = geographic information systems; ICT = information and communication technology.

Precision agriculture

Precision agriculture has the potential to improve productivity and generate environmental and health benefits. Precision agriculture technology effectively combines data-intensive analytics with data-guided farming equipment (box 5.1), generating environmental benefits such as water savings and reduced nutrient runoff and pesticide use (Mondal and Basu 2009). Benefits to farmers include savings in labor and input costs (seed, fertilizer, pesticides, water), optimized timing of crop management practices, ability to give insight into crop variability over large acreage, and reduced exposure to pesticides.

BOX 5.1

The future of farming: Artificial intelligence and precision agriculture are fast changing the technology landscape

Machines called "agribots," from a variety of companies, are appearing in the agricultural fields in many shapes and sizes. The development of agribots is driven by artificial intelligence. These electrically powered devices can do a variety of agricultural tasks, such as monitoring of crops (nutrient status and presence of weeds, pathogens, or pests), weeding (electrocution), spraying and micro-dosing nutrients or pesticides, hoeing, and harvesting. Self-contained agribots will have to compete with systems towed by smart tractors. Most modern tractors and combine harvesters can steer themselves across fields using satellite positioning and other sensors. Some tractors use digital maps of crops obtained by satellites and drones to highlight the places that require fertilizer or pesticides (*Economist* 2020).

Asia's wealthier nations are advancing the Internet of Things (IoT) and automation in field monitoring and precision agriculture. In 2016, Japan opened the world's first robot farm. The Singaporean firm Garuda Robotics is providing drones to Southeast Asian farmers. In the Republic of Korea, the government began testing a "smart farm village" in Sejong City in 2015, providing farmers with a suite of smart agriculture tools including remote sensing and automated controls, all connected to smartphones, resulting in a 23 percent increase in efficiency. Malaysia is also making great strides—the government included agriculture as part of the national 2015 IoT plan, incorporating a pilot project that applied the IoT to aquaculture traceability. Malaysia's information and communication technology research and development center is also conducting trials of sensor technology to help plan the timing of oil palm pollination (Green 2018).

China is also experimenting with precision farming. The modern agriculture project in Hubei province uses the BeiDou Navigation Satellite System, which combines high-precision positioning technology with sensor technology to realize accurate monitoring of soil moisture, farm machinery autopilot control, and direct seed precision planting. Another example is the intelligent rice bud production system in Heilongjiang province, which conducts real-time data collection through temperature and moisture sensors in greenhouses to achieve intelligent micro-spraying and electric shutter ventilation control (ADB 2018).

Precision technology and field-monitoring tools must be adapted to smallholder contexts to have an impact. Most field-level precision agriculture innovations are not about agribots or automation but about less advanced and less costly applications. The most appropriate tools for smallholders are often singular, low-cost tools, such as chlorophyll meters, although smallholders involved in cooperatives can make use of larger precision agriculture packages (Giovannucci et al. 2012; Ortolani and Bella 2015; Teng 2017). Other notable low-tech innovations in the region include kits for digital soil testing, smallholder algae production (Feed the Future 2017), and solar-powered irrigation (Arizona State University 2017). The Philippines has been using unmanned drones equipped with navigation and photographic technology to identify land vulnerable to natural disasters, and the country's space satellite program included the launch of Diwata-1,

continued

Box 5.1, *continued*

designed to provide forecasts and weather monitoring and to survey farmland. Thailand's Mahidol University has also used drones for monitoring crops using multispectral, hyperspectral, thermal, and LIDAR (light detection and ranging) imaging (Green 2018).

Many precision innovations require little capital and can be developed by start-ups and entrepreneurs. Ag-tech communities in Asia are developing a number of tools, satellites, drones, weather-tracking tools, and remote sensors. Regional examples include the following:

- Vietnam-based MimosaTEK has developed a cloud-based system that allows farmers to control and manage their farms using sensors. It produces alerts in response to challenges and can execute remote irrigation by mobile phone.
- HelloTractor has developed a two-wheel smart tractor with a global positioning system antenna, allowing smallholders to request and pay for tractor services through text messages on a just-in-time basis. Farmers have seen yield increases of as much as 200 percent since its launch in 2014. HelloTractor is prominent in Africa, and has been introduced in India, Pakistan, and Thailand since 2019 (HelloTractor 2019).
- Indonesia-based eFishery developed a smart fish feeder that automatically dispenses feed when fish are hungry. In a region with high fish consumption, and with feed representing 80 percent of production costs, innovations such as this (reducing feed costs by 20 percent) could have profound benefits (Green 2018).

The transition to greater use of precision agriculture among smallholders requires attention to structural issues (for example, land rights and lot size), infrastructure, and affordability (Green 2018). East Asian countries are at different stages of development or of adoption of precision agriculture. Japan, the Republic of Korea, and Singapore are in the lead and adopting sophisticated and automated farming, such as agribots that manage most field work, while China, Malaysia, and Thailand follow at some distance. The challenge for developing East Asia will be to adapt precision agriculture technologies to the region's agro-ecological and small-farm realities, where cost, ICT connectivity, and skills also play a role (box 5.1).

E-extension and other e-services

Mobile-based extension and other e-services have increased smallholders' productivity and resilience to climate risks.[3] ICT applications in agricultural extension, weather and price information, banking, and to some extent insurance[4] have been successfully promoted in several regions since about 2005 (World Bank 2012, 2017a). Increasing access to mobile phones has enabled delivery of timely and customized information at scale, contributing to timely farming and market decisions and yield increases.[5] Improved access to timely information is particularly important in East Asia, where small farms operate a significant share of farmland—about 70–80 percent of farms are smaller than two hectares (Cassou, Jaffee, and Ru 2017).

Most developing East Asian countries have tried to optimize the outreach and coverage of extension with diverse mobile-based applications. Most e-extension services are subnational, nimble, and managed by nonpublic actors, such as the Vinaphone-managed "farm assistant" program in Vietnam (box 5.2). China has, however, attempted to streamline e-extension as part of its broader public sector

BOX 5.2

Vinaphone-managed mobile-based farm assistant

Vinaphone, a Vietnamese state telecom company, launched a service that turns mobile phones into "farm assistants," providing agricultural information services (such as weather forecasts, plant disease alerts, guidelines on government policies, and advice on plant diseases) through a low-cost monthly package, delivered through social media. Similar services exist in Indonesia, through Villages.

Source: Green 2018.

extension system by wide-scale implementation of online agriculture information platforms and farmer-friendly information stations. Details on ICT use for smallholder services in East Asia are provided in appendix E (box E.1).

Blockchain technology for productivity

Blockchain technology[6] offers new opportunities to improve agricultural productivity, along with inclusivity, traceability, and the means to monitor compliance with environmental regulations, green or carbon bonds, and food safety (box 5.3). Initially born out of a need for a more decentralized financial system, blockchain applications are finding multiple innovative uses in financial, manufacturing, health, energy, and government sectors, as well as in the agriculture sector. Most applications in agriculture are in a pilot stage or in the early commercialization stage (appendix E) (FAO 2019a). Blockchain applications are also being pursued to improve transparency and traceability in agricultural supply chains.

Blockchain promises to deliver a transparent, decentralized, secure transaction process and may reduce transaction costs. However, blockchain technology is not yet mainstream, and commercial applications take time to reach the market. China is experimenting with blockchain applications in food traceability, and many other East Asian countries have high aspirations to develop this area as well (for instance, in Vietnam, box 6.6 in chapter 6). Implementation of blockchain-based applications still suffers from traditional challenges such as lack of or poor infrastructure, failures of interoperability, and other technology issues. Regulators around the world are actively trying to define guidelines for distributed ledger usage, suggesting that this technology can bring about a change in traditional processes if key building blocks needed to sustain the solutions are implemented (FAO 2019a).

Agricultural productivity drawing on biotechnology

High-tech scientific research, including biotechnology, is essential for improving the resilience of agri-food systems to changing climate, diseases, pests, and degradation of natural resources. Despite advancements in ICT-based solutions, high-tech science still has a role to play. The current challenges with climate risk and environmental degradation have created a need for more advanced scientific endeavors in inputs, especially the genomic basis of crops, as part of a menu of climate-smart agriculture (CSA) approaches.[7] A diverse range of

BOX 5.3

Blockchain applications in the agri-food system

Insurance. The FARMS pilot in Kenya is enabled by a blockchain-based virtual currency platform integrated with remote sensing (satellite) data and mobile money solutions, which ensures transparent, secure transactions and "earmarking" of funds, automated payment, and information dashboards. Farmers set aside money by buying virtual currency "drought coins" or "drought vouchers" that are kept in their personal COIN22 mobile wallet account. When farmers want to withdraw funds, they redeem their drought coins or vouchers. The actual money flows into a trusted bank account (risk pool), and through full systems integration all transactions are in real time.

Cash transfers. The World Food Programme is testing blockchain as part of its "Building Blocks" pilot, to make cash transfers more efficient, secure, and transparent. The World Food Programme is also using blockchain to deliver food assistance more effectively to Syrian refugees in Jordan.

Land registration. The United Nations Development Programme in India is working with partners to make land registry more reliable. At a high level, this project will capture and permanently record each transaction throughout the sale of a property.

Fisheries sector. The World Wide Fund for Nature in New Zealand is piloting a project to track and deter illegal, unreported, and unregulated fishing and human rights abuse in the Pacific Islands' tuna industry. The value chain from fish in the vessel to the supermarket can be tracked.

AgUnity—a transparent cooperative system in Kenya and Papua New Guinea. AgUnity has developed a solution that is providing a pathway to financial inclusion for the world's poorest farmers. The AgUnity app is a simple mobile service that helps small farmers plan, trade, and track everyday transactions. This is a way for farmers to cooperate, store value, get paid upon delivery of produce, hire equipment, save money, and easily buy products and services.

Carbon credits. Companies such as Poseidon use a blockchain-based system to track an individual's or company's carbon footprint and then provide opportunities to offset it. IBM works with Veridium to tokenize carbon credits that are verified by third parties. These are then used to provide incentives to companies to be more environmentally friendly and to offset their carbon footprints.

Source: Adapted from FAO 2019a.

biotechnological applications are in use or under development in the crop sector in the region. These applications range from the less advanced, such as biopesticides, biofertilizers, and tissue-culture techniques, to technically advanced applications such as genetic modification and genome editing of crops (FAO 2019c).

Genetically modified crops

Asia has already benefited from biotechnology, including seed and crops that are resistant to flooding, drought, or temperature changes (box 5.4) or pests (Fuglie et al. 2020; Green 2018; ISAAA 2019). Globally, Brazil, India, and the United States are frontrunners in the development and adoption of genetically modified (GM) crops. GM crops have also been approved in several developing East Asian countries for commercial production (table D.1 in appendix D), and a wide range of new food and raw material crops and traits are either at the experimental stage or in field trials.[8] China has led the development of GM crops in East Asia (box 5.5). Attitudes among farmers are generally positive because GM crops can

BOX 5.4

Genetic modification in rice and adaptation to climate change

The International Rice Research Institute and its partners have developed new generations of rice that can tolerate submergence, drought, heat, and salinity. More than 100,000 rice varieties have been developed, delivering benefits valued at $1.46 billion per year and boosting rice yields by an average of 11.2 percent annually in Southeast Asia (RSIS 2014; UNSDSN 2013). India has also developed rice that can tolerate 15 days of flooding. Singapore's Temasek Holdings has developed Temasek rice, which can withstand extreme weather (Green 2018).

BOX 5.5

Agricultural research and development and production of genetically modified crops in China

The Chinese government identifies biotechnology as a strategic industry, but low public acceptance has continued to impede adoption by farmers. Agricultural biotechnology research started in the early 1970s. By the late 1970s, China began research and development of genetic engineering. Genetically modified (GM) crops in China were developed in the mid-1980s, and virus-resistant tobacco was the first commercialized crop in the early 1990s. Although China has commercialized six GM plants since 1997 (cotton, papaya, petunia, poplar, sweet pepper, and tomato), only papaya and cotton are in production today because of difficulties in bringing the products to commercialization. An important roadblock is public concern about the safety of GM crops. The government is also concerned that China's food biotechnology will be controlled by multinational companies, thus affecting the country's food security (Cai et al. 2017). In 2014, China was the sixth-largest producer of GM crops worldwide. Since 2016, China has aimed to promote the commercialization of key products, including the new generation *Bacillus thuringiensis* (Bt) cotton, Bt corn, and herbicide-tolerant soybeans.

China has gradually developed its biosafety regulatory system, including enforcement. China established a biosafety regulatory system in the mid-1990s and strengthened it in the early 2000s. The discovery of unapproved transgenic crops and unlabeled derived products indicated that enforcement of the biosafety regulatory system needed improvement (Karplus and Deng 2006). The National Biosafety Committee developed guidelines for biosafety assessment (environmental and food safety) to streamline the processes. The committee consists of 44 experts with diverse backgrounds from different Chinese ministries, research institutions, and universities. In 2016, China also revised the biosafety regulation on genetically modified organisms. The amendments remove timelines for approvals, extend the National Biosafety Committee's term from three to five years, and emphasize that entities engaging in research and experiments on genetically modified organisms are accountable for safety management.

The lack of a clear path for commercialization of major GM varieties has limited incentives for local seed companies to invest in biotechnology. It has also encouraged public labs to focus on basic research rather than develop commercially viable seeds. Inconsistent protection of intellectual property rights and the fragmented nature of China's seed industry further discourage private sector investment in biotechnology.

increase yields and profitability and reduce potential adverse environmental and health effects associated with pesticide use. In China, for example, more than 90 percent of cotton growers have adopted GM varieties. However, urban consumers are still resistant to the adoption of GM technologies for food crops. The government is sensitive to this resistance and sees wider adoption of GM technology not as a technical issue but as a public relations and awareness issue.

New breeding techniques, including genome editing

A new generation of techniques known as new breeding techniques can help the East Asia region develop more resilient and nutritious crops. The new breeding techniques include a variety of techniques that introduce genetic mutations indistinguishable from the processes of natural breeding. Genome editing (GE) has already been used in agriculture in China, Japan, and Korea. Many activities are at the experimental stage (FAO 2019c). Current projects and initiatives include research into commercially and nutritionally important crops such as rice and wheat (box 5.6). China has invested significantly in GE as part of a wider, technology-based push to improve agricultural output, and is about to become the global leader. Differences in regulation between the European Union and other countries such as the United States have created potential barriers to the use of GE, however. In 2018, the European Union declared that GE crops should be subject to the same stringent regulations as conventional GM organisms (Callaway 2018). Elsewhere, as in the United States, decrees and enforcement ordinances are generally encouraging of GE (FAO 2019c; Green 2018).

BOX 5.6

New breeding techniques

New breeding techniques (NBTs) include techniques such as genome editing (GE), reverse breeding, and agro-infiltration, with GE attracting the most attention. Whereas genetic modification involves the artificial transfer of genes between organisms that are not bred, or the introduction of genes from outside an organism's genome, GE methodologies (clustered regularly interspaced short palindromic repeats [CRISPR] and transcription activator-like effector nucleases [TALEN]) introduce genetic mutations that can be indistinguishable from those found in natural breeding. CRISPR has become the preferred GE technique because of its simplicity and efficiency.

In agriculture, the NBT trend seems positive because it can help strengthen resistance to pathogens in crops such as rice and tomatoes, and prolong shelf life. In the United States, GE crops are already on the market, including a browning-resistant mushroom, a waxy maize that produces higher starch, and an improved storage potato. Berkeley-based Caribou Biosciences is also using GE to create drought-resistant corn and wheat.

China has invested in GE technology since 2014. It has at least 20 research groups across the country dedicated to GE use in agriculture. Chinese scientists have used GE technology to create soybean mutants that can adapt to low altitude areas, paving the way for the breeding of new soybean varieties. The soybean mutants exhibited late flowering, improved height, and an increased number of pods, providing a basis for the breeding of soybean varieties that grow well in low altitude regions. Many activities are at the experimental stage. Current projects and initiatives include research on commercially and nutritionally important crops such as rice and wheat (resistance to powdery mildew), and cold-resistant, lean-body-mass pigs. The first GE product, high oleic soybean oil, was successfully commercialized without the regulations applied to genetically modified crops.

Sources: Cohen 2019; FAO 2019c; Green 2018.

Microbiomes

Microbiomes have the potential to improve crop resilience and reduce greenhouse gas (GHG) emissions.[9] The plant microbiome refers to the environment of microorganisms in and around the roots, in the soil, on the leaves, and within the plant itself. When applied directly to the surface of seeds and to plants, microbiome technologies can complement or replace agricultural fertilizers and pesticides. The results are impressive: higher yielding, healthier crops that are more resistant to drought and are tolerant of lower levels of nitrogen application, higher temperatures, saline soils, and pests. Companies are now trying to identify the organisms that are most beneficial in different environments and to create products based on them. For example, the company Indigo has brought to market microbial products for corn, wheat, soybeans, and rice (WEF 2018). For further details, see appendix E.

Livestock feed additives

Feed additives for improved growth and health of livestock are another area of intense innovation. The private sector is the driver, responding to the swiftly growing market segment globally. The rapid sales growth of feed additives in the East Asia and Pacific Region is mainly attributed to the growing animal protein production and animal feed industry (3 percent growth between 2017 and 2018) (Meticulous Research 2020). Innovations in feed ingredients have the potential to support improvements in efficiency and profitability of livestock production, reduce the environmental footprint, and improve animal health and welfare. For instance, interest in identifying alternatives to antibiotic use is driving some of the research (Liu et al. 2018). Probiotic feed additives (FAO 2016; Mitchell 2019), multi-carbohydrase feed enzyme technology for nonruminants, nucleotides, and yeast-based supplements (Bedford 2018; Liu et al. 2018) are examples that have shown potential to improve animal immune systems, regulate gut microbiota, and reduce the negative impacts of weaning and other environmental challenges.

Producing food off farms and from alternative sources

The changing preferences of urban consumers have increased interest in food produced off farms and from alternative sources. Consumers increasingly prefer safe, fresh, and more nutritious food produced closer to where it is consumed. Pressures related to rising food demand, the continuing loss of good agricultural land and farmers to other uses, and the demand for a reduced environmental footprint have further stimulated this interest.

Urban agriculture

Urban agriculture—the growing of plants and the raising of livestock within cities and peri-urban areas (RUAF Foundation 2018)—shows increasing potential for delivering safe food with limited environmental costs. Urban farming is not new; it has been practiced in semiurban areas and in backyards in Asian cities for ages. However, it has gained in importance in parallel with global megatrends and evolving technology (such as vertical farming, integrated sensors, controlled lighting). According to estimates for 2030, urban farming may be valued at $40 billion, half of it in the Asia region (Ecosperity 2018a).[10] Urban farming has many benefits, including improved food security and freshness of food, resilience to disruptions in the food chain, a reduced environmental footprint through efficient water and nutrient use, and lower costs through reduced use of inputs, transportation, and storage. For example, indoor vertical farming in Japan can save up to 99 percent of water compared with outdoor field farms (Ecosperity 2018a). Livestock production in peri-urban areas can, however, pose an increased risk of emerging infectious diseases (World Bank 2010).

Urban farming technology has reached a point at which the physical constraints, such as land, water, and weather conditions, can be overcome to some extent. Urban farming can be practiced in fields and backyards, but increasingly it takes place on rooftops and building walls and in deserted industrial buildings and balconies. Vertical farming (food grown in vertical layers), integrated sensors, hydroponics (plants growing in a nutrient solution), and controlled lighting allow cities to use small plots of land to grow crops faster (Ecosperity 2018a). Although vertical farming is still small scale worldwide, the COVID-19 pandemic has stimulated interest and investment in vertical farming globally (Financial Times 2020a). Urban farming can also entail low-tech applications (for example, growing kits and information packs).

The highly urbanized, technology-savvy nations of Japan and Singapore are the frontrunners in urban farming, but developing East Asian countries follow. Several cities, including Beijing, Shanghai, and Tokyo, are all seeing upticks in urban farming. There are key differences regarding urban agriculture across East Asia in the level of adoption and commercialization as well as the technology used. Japan and Singapore (box 5.7) have invested heavily in automated production, hydroponics, and vertical farming. Singapore recently announced that it aims to increase its food self-sufficiency from the current 10 percent to 30 percent by 2030 in response to disruptions in food imports caused by COVID-19 (PwC and FIA 2020). Most of this additional food is expected to come from its urban farming systems. Rapidly urbanizing China is also embracing urban farming (box 5.7).

Barriers to urban farming are context-specific. Barriers typically include land scarcity, high up-front capital and energy costs, uncertain regulations, and entrenched consumer behavior (box 5.7). Citizens might lack access to the financial or natural resources (such as land, seeds, water, and fertilizers) and the knowledge needed to set up production. Middle-income East Asia can, however, use a range of levers to encourage urban agriculture, from streamlining regulations to providing technical assistance and funding (Green 2018).

Protected agriculture

Protected farming for high-value vegetables and fruit has also gained ground, often in association with urban farming and drip irrigation. Asia accounts for 75 percent of the agricultural area under protected farming worldwide (Lamont 2009). Protected agriculture is typically practiced on farms but is a common practice in urban farming as well. It covers a range of practices from low-cost polytunnels made of recyclable materials, which rely mostly on natural solar energy input, to very expensive and sophisticated plant factories, which exclude natural solar energy and rely almost exclusively on artificial energy input (Kang et al. 2013). Plastic use is common in protected agriculture (appendix A). For example, plastic film, tunnels, and irrigation pipes have contributed to enhanced crop growth and product quality and have reduced water

BOX 5.7

Urban agriculture in East Asia's agri-food systems

Urban agriculture is a crucial component of Japan's economy and agri-food system. Up to one-third of Japan's agriculture output is generated in urban areas, and urban farmers account for a quarter of farm households (Moreno-Peñaranda 2011). Urban agriculture is highly commercialized, and farms have grown from an average of 641 hectares in 2005 to 877 hectares in 2015 (Sim 2018). Japan is also a hotbed for technology advances and is a pioneer in information and communication technology–enabled indoor crop production, use of drones for harvesting, and innovative green applications such as using edible crops to provide insulation for buildings (Ecosperity 2018a). Japan is also an Asian frontrunner in vertical farming, with Spread, Fujitsu, and Aerofarms all pursuing hydroponic (soil-less) and vertical agriculture. Spread, which has been active in vertical agriculture since 2006, now produces more than 20,000 heads of lettuce a day, shipped to more than 2,000 supermarkets (Goedde, Horii, and Sanghvi 2015). Fujitsu's entry into high-tech agriculture is seen as an effort to deploy greater technology research and development in the food sector. The same is true of Toshiba, which has converted one facility into a lettuce and salad factory (Green 2018).

Singapore has been leveraging technological advances in vertical farming using methods such as hydroponics and aeroponics. The number of vertical farms has grown from just one in 2012 to seven in 2016, producing a range of produce from vegetables to aquaculture (Singh 2016). For example, the Apollo Aquaculture Group has created a local "high-rise" seafood farming project that produces six times more than a traditional aquaculture project, and everything is remotely controlled and carefully managed, including the amount of fish feed dispensed (Ecosperity 2018a).

The Urban Redevelopment Authority in Singapore has lowered the barriers for urban farming by allowing urban farms and communal gardens on rooftops to contribute to landscape replacement requirements. In addition, longer urban farm leases (20 years instead of 10 years) can also encourage greater uptake of expensive farming technologies. Urban farmers can leverage Singapore's Agri-Food and Veterinary Authority's $47 million Agriculture Productivity Fund to defray high adoption costs.

Rapid urbanization drives urban farming in China. Environmental factors such as depleted arable land and water contamination along with urbanization make it more important for cities to engage in urban agriculture (Bloomberg 2017). For instance, the number of greenhouse companies has grown from 5 in the 1980s to about 400 in 2010 (Smart Agriculture Analytics 2015). Beijing was one of the first cities to integrate urban agriculture into its overall development strategy by developing five "agro-parks." The city also attempted to institutionalize urban agriculture by measuring and documenting its economic, social, and environmental impacts in official records beginning in 2010. The Shanghai government has launched programs to create a sustainable system of urban farming, including quality control systems as well as government-funded campaigns to promote food safety and consumer acceptance. At the national level, about 40 research institutes are working on solutions that will boost efficiencies in vertical and indoor farming.

and pesticide use (providing improved resilience to climatic and weather risks) (Schuttelaar and Partners 2019). However, plastic use in agriculture has also contributed to an increase in marine debris, given that much of the agri-food plastic ends up in oceans through riverways (box 5.8) (World Bank 2019c). Marine debris, originating from many sources, is a serious threat to the environment, the economy, and health. The annual global damage to marine ecosystems caused by plastics is estimated to be at least $13 billion per year (World Bank 2019c).

BOX 5.8

The agri-food sector and increases in marine plastic debris

Plastic generation worldwide has increased at alarming rates, with Asia accounting for 50.1 percent of the worldwide total of 348 million tons in 2017. Packaging is the greatest generator of plastic use (36 percent) (UNEP 2018). Increased use and poor waste disposal or recycling practices have resulted in marine debris that adversely affects ecosystems, public health, and maritime country economies, particularly in the fisheries, aquaculture, coastal tourism, commercial shipping, and agriculture sectors, and degrades ocean natural systems. There are currently 150 million tons of plastics in the oceans, and the number could reach 250 million tons in less than 10 years if current trends continue (Jambeck et al. 2015).

Plastic use in the agri-food sector has also increased steadily by 20 percent per year between 1992 and 2002. By 2002, 586,300 hectares worldwide were under protected crops (greenhouses and high tunnels), of which the Asia region accounted for 75 percent (Espi et al. 2006; Lamont 2009). China's plastic film use increased from 1,250,000 hectares in 2002 to 18 million hectares in 2014 (Espi et al. 2006; Liu, He, and Yan 2014). The amount of plastics used in agriculture worldwide was estimated at 2 percent of global production in 2010 (Vox et al. 2016). Changes in the retail sector and demand for convenience, food safety, and freshness and a longer shelf-life as well as smaller product sizes have also contributed to an increase in plastic use in the food and beverage sector (World Bank 2019b). For instance, in Indonesia, food and beverage already account for about 70 percent of plastic packaging (World Bank 2020). Recent information points to a significant increase in plastic use as a response to changes induced by COVID-19 (coronavirus) in the food chain (and economywide). Appendix A and table A.1 offer additional data on plastic use in agriculture and regional generation and management of plastic.

Plastic use in the agri-food sector has contributed to productivity, food safety, and resilience to climate change, but there are also significant trade-offs that warrant innovation in reduced usage, reuse and recycling, alternative and biodegradable materials, and agricultural production practices less reliant on plastic. The private sector is already developing innovative alternatives to plastic packaging to capture a share of the vast food and beverage packaging market (50–60 percent (European Bioplastics 2018; Korhonen, Koskivaara, and Toppinen, n.d). For instance, cardboard packages and fiber-based packaging materials with plastic-like features are expected to reach a larger segment of the market in the future.

INNOVATIONS ACROSS THE FOOD CHAIN IN TRACING, PACKAGING, DISTRIBUTION, AND CONSUMPTION

Greater efficiency throughout the agri-food value chain can promote the broader goal of sustainable food production. The lack of traceability of food across different stages of the value chain can create issues not only for price transparency but also for food safety and food fraud. Public sector governments can enhance traceability and food safety by enhancing communication channels, ensuring adequate regulatory frameworks, promoting certification for global food safety standards, and introducing new diagnostic tools to promote price transparency (Ecosperity 2018a; World Bank 2019b). There is also considerable room for private sector innovation. Technologies can optimize supply and demand matching (for example, e-commerce), and companies are already exploring internet-based food and grocery delivery, helping to create more densely networked and nimble food delivery systems. Selected innovations affecting the food chain are explored in this section.

Blockchain technology for traceability

Blockchain technology in the value chain can help track food as it moves along the supply chain, drive food supply transparency, and increase responsiveness to food safety issues. The technology would enable specific products to be traced at any time, allowing contaminated products to be traced easily and quickly. This ability would ease product recall while reducing food waste because safe foods can remain on the shelf. AgriDigital is one of the global pioneers in the use of blockchain for managing agriculture value chains (box 5.9). Others include Covantis, GrainChain, and Agriota E-Marketplace. In late 2020, Swiss-based blockchain commodities marketplace Cerealia went into production, initially targeting the grain trade in the Black Sea area (Ledger Insights 2020). For challenges, see the subsection titled "Blockchain technology for productivity" earlier in this chapter.

Food sensing technology

Other innovations improving traceability, quality, and food safety include a wide range of sensing technologies. These innovations include antifraud solutions, diagnostic technology, antibacterial proteins, mobile-based tracking systems, and sensors for the nondestructive analysis of food (box 5.10) (ADB 2018; Ecosperity 2018a; WEF 2018). The technologies are often applied at the processing and retail levels.

BOX 5.9

AgriDigital: A pioneer in the use of blockchain in the agricultural value chain

AgriDigital has a cloud-based commodity management solution for the global grains industry, connecting grain farmers, buyers, site operators, and financiers through a single platform, and allowing them to contract, deliver, and make payments securely and in real time. The AgriDigital platform is also blockchain enabled. With the launch of a commercial blockchain protocol for agriculture, the AgriDigital platform operates as the primary application layer for users to interact with the blockchain. The company has developed a library of smart contracts operating on the blockchain protocol, allowing users to trade, finance, and trace agri-commodities. At the core of AgriDigital's solution is the creation of digital assets. Using digital assets, AgriDigital brings together the trade, finance, and data flows that are often disparate in traditional, paper-based agri-supply chains. This provides users with a more informed and robust view of their assets and the supply chain, and improves liquidity, transparency, and security for all supply chain participants, including farmers, traders, producers, financiers, and consumers.

AgriDigital's SaaS34 platform has a network of 1,300 active grain supply chain users with a broader network of more than 4,500 total users. To date, more than 1.6 million metric tons of grain have been transacted by these participants on the AgriDigital platform (since November 2016). It is expanding rapidly into the global grains industry and across commodities with initial trials underway in the rice and cotton industries.

Although based in Australia, AgriDigital has expanded overseas, including to North America.

Source: ADB 2018.

BOX 5.10

Recent innovations to improve product traceability and food safety testing

Antifraud solutions. Demand has increased for anti-counterfeit packaging, which consists of smart packaging such as radio-frequency identification technology, near field communications, and holograms to enhance food security. Radio-frequency identification tags can be used by companies to track food products from the factory to the supermarket.

Food diagnostic technology. Food safety testing services cover a wide range of testing methods for different strains of pathogens. For instance, polymerase-chain-reaction testing identifies major pathogens such as *E. coli*. Service providers can also check for allergens on behalf of food companies. These technologies are used mainly in the European Union and North America.

Antibacterial proteins. Certain proteins have antibacterial properties that neutralize produced food that is prone to infection. Such applications can target specific strains of bacteria and improve food safety at the mass market level. AvidBiotics is currently developing Purocin, which has food safety applications including against salmonella.

Mobile-based system for tracking products. The Shouguang Municipal Bureau of Agriculture in China has created a quality tracing system for vegetables, which farmers can use for free. Using a mobile app to scan the quick response or QR code, customers can access information such as cultivation base, sampling time, results of pesticide concentration, test planting (pruning, splitting, and watering), harvest, and sales transaction data (ADB 2018).

Sensors, such as near-infrared spectrometers and hyperspectral imaging. These sensors are increasingly being used to conduct nondestructive analysis of food. This application combines spectroscopy with computer vision. Images are analyzed via the cloud using machine learning and imaging-processing algorithms to interpret the data, resulting in actionable information such as quality, safety, and authenticity of food. The information generated from scanning technologies can determine the freshness of food and could replace the need for sell-by and use-by dates. If sensing technologies could reach 30–50 percent of the consumers in developed markets, domestic food waste could decline by 10–20 million tons by 2030. At present, sensors are not used on a large scale (WEF 2018).

Alternative packaging

The food and beverage packaging industry is undergoing a revolution to respond to consumers' and society's needs. Apart from catching the eye of consumers, packaging is expected to protect the product, prolong the shelf-life of fresh produce (for example, through use of modified atmosphere packaging), and satisfy various requirements for traceability, sustainability, and reduction in food loss and waste (FLW) (see box 5.10). Over the years, plastic use in food and beverage has increased in parallel with urbanization and consumer demand for convenience, food safety, freshness, and e-commerce (World Bank 2019a). Consumer concern about plastics has grown considerably, however. Although using recyclable materials for packaging is one way to address sustainability concerns, some companies ensure their packaging is made from biodegradable or compostable materials[11] (Burrellon 2019; European BioPlastics 2018; Kemira 2020). Advanced technologies, such as ultrasonic sealing, can also be used to minimize seal sizes, which in turn reduces material use. Today, alternatives to plastic packaging (for example, fiber and plant-based materials) and bioplastics[12] (for example, bio-based, biodegradables, and fiber composites) are already in use or in various stages of development in many high-income countries (HICs) (European BioPlastics 2018; Kemira 2020; Korhonen, Koskivaara, and Toppinen, n.d).

In addition, a substantial increase is expected in edible packaging made of materials such as rice paper, seaweed, or potato. Packaging can also offer an interactive experience for consumers; interactive augmented-reality packaging has grown 120 percent in recent years. Interest in transparency extends into food packaging not only in the information provided on the package but also in the material used. The amount of clear packaging made from translucent materials has increased recently, allowing consumers to see exactly what they are purchasing in the product. With the expansion of e-commerce, packaging innovation should also include lighter-weight yet durable elements that stand up to shipping demands (Burrellon 2019; Keating 2019). The cost and quality features (for example, durability, resistance to temperature changes, moisture retention, recycling) along with sustainability are important areas for continued effort and innovation.

Reducing food loss and waste

Reducing FLW is an integral part of improving food security, enhancing resource use, and reducing GHG emissions (arising from logistics, energy use, and wastelands). Total FLW is estimated to be about $1 trillion in 2016, of which roughly $680 billion is lost in HICs. About 40 percent of this FLW occurs at the retail and consumer levels. About $310 billion worth of FLW occurs in developing low- and middle-income countries, where more than 40 percent of FLW occurs at the postharvest and processing levels (FAO 2017). In China alone, more than $32 billion worth of food was thrown away in 2013 (Zhou 2013).[13] Most innovation in FLW takes place in HICs. In East Asia, Japan, Korea, and Singapore as well as China (for example, the Clean Plate Campaign) and Malaysia have embraced FLW solutions (Austrade 2019; Ecosperity 2018a). FLW has also received considerable international attention; several initiatives on FLW have been launched.[14]

FLW may be reduced through individual innovations as well as through regional and system-level approaches. FLW is prominent at different points along the farm-to-fork continuum. Research and weather systems applications, including biotechnology-based crop breeding and e-weather services, together with cold storage solutions can help address farm-level production, harvest, and postharvest-related losses (see the section titled "The transformative role of technology in agricultural productivity" earlier in this chapter). Digital solutions that help match consumer demand with supply (for example, e-commerce, discussed later in this section), new technologies such as smart labels and sensors for analysis of food (box 5.10), and innovations that extend the shelf-life of fresh products (such as biotechnology-based crop breeding and controlled atmosphere packages) may also reduce retail- and consumer-level FLW. Data systems that better manage production and logistics processes can contribute to reduced FLW. For crops that cross borders, electronic customs processes and efficient trade corridor systems and transport connectivity initiatives, such as the Greater Mekong Sub-Region project and the East ASEAN (Association of Southeast Asian Nations) Growth Area, which has food as one of its six strategic pillars, can smooth logistics (Green 2018).

The concept of the circular economy has become increasingly important in the design of production-consumption systems that limit or terminate waste and pollution, keep products and materials in use, and regenerate natural systems (EMF, n.d.). In the context of the agri-food system, food waste may be turned into, for example, animal feed, biogas, fertilizers, and building materials.

Many circular economy agri-food system activities are, however, in the early stages. For example, AgroCycle is a three-year project with 25 partners from China, the European Union, and Hong Kong SAR, China. It aims to further develop, demonstrate, and validate novel processes, practices, and products for the sustainable use of agricultural waste, co-products, and by-products (Toop et al. 2017). A traditional example of a circular economy is the VAC (orchard, pond, animal in Vietnamese) integrated system in Vietnam (Nhan et al. 2005).

The challenge in developing East Asia is to identify FLW solutions that accommodate smallholders. A significant concern is FLW occurring in the value chains consisting of myriad smallholders with poor transport and storage systems. Solutions that improve storage conditions, the shelf-life of perishables, and logistics (for example, drones, collection centers) are needed. Novel cold storage solutions, such as stand-alone solar-powered storage, may be an option in off-grid areas or areas with unreliable electricity (Economic Consulting Associates Limited 2020). For instance, Ecozen in India developed micro cold storage, a solar-powered cold storage system. After two years of use, small-scale farmers can see an increase in profits of more than 40 percent (see appendix E) (WEF 2018). Once urbanization accelerates, many FLW innovations are expected to take hold to a greater extent in other countries.

E-commerce

E-commerce has grown significantly globally and regionally[15] and has received an unexpected stimulus in response to COVID-19 impacts on the food chain and consumers[16] (chapter 2). Overall, e-commerce in food and agricultural has been developing slowly (box 5.11). One of the challenges is that e-commerce-based food delivery is more viable in urban areas, and many of Asia's residents still live in rural areas (Green 2018). E-commerce may not necessarily benefit smallholders—economic viability depends on large retailers working with delivery and logistics entities, rather than small farmers selling produce directly to consumers on the internet. Indeed, e-commerce has been particularly weak when it comes to connecting smallholders who produce perishable foods because the transport and logistics costs of getting those foods to consumers are not economical on a just-in-time basis (ADB 2018). Purchase of nonperishable farm inputs may lend itself better to e-commerce, however. Rural agriculture e-commerce has also received significant attention as a response to the COVID-19 crisis. Many e-commerce enterprises have mobilized resources in procurement, logistics, operations, and marketing to maximize benefits to both farmers and consumers.[17]

FOOD CONSUMPTION AND NUTRITION: FROM BASIC SUSTENANCE TO PERSONALIZED NUTRITION

Innovation and technology in food is an emerging area of exploration and investment in East Asia. It is driven by consumers who place increasing emphasis on convenience—greater consumption of processed, instant, and fast foods, but also of fresh food[18] (Ecosperity 2018a)—and on reducing the environmental footprint of food. It requires rethinking food intake by exploring new sources of nutrition with a lower environmental footprint (for example, insects, artificial meat, vertical farming) and greater health benefits (such as food fortification, functional foods, fresh food, "clean food," reformulated and "free-from" foods, nutrigenetics). Another important driver of nutrition innovation is the region's

BOX 5.11

Emerging but struggling food e-commerce

Because of high population densities in urban areas, the East Asia region is well-suited to food delivery services via e-commerce. Large multinational companies, such as Amazon and Walmart, have demonstrated interest in fresh food commerce in the region. Asian e-commerce players could, however, dominate the food delivery business. China has several ongoing initiatives, and Japanese messaging service Line has taken steps into perishable and nonperishable sales and delivery in Southeast Asia. Companies have launched similar businesses in Indonesia, Malaysia, and Thailand (Green 2018).

Evidence suggests that e-commerce platforms are being used to connect producers directly to consumers. DHL is currently working with Thailand's Ministry of Commerce to deploy e-commerce and logistics expertise to connect Thai farmers to e-commerce platforms (Green 2018). China has implemented e-commerce for the agriculture sector (via Taobao.com and Pinduoduo). The government's Rural E-Commerce Demonstration Program seeks to promote e-commerce in rural areas by establishing and improving rural e-commerce public service, fostering rural e-commerce supply chains, promoting connectivity between agriculture and commerce, and enhancing e-commerce training (World Bank 2019a). The rural e-commerce industry chain covers a range of activities, from the production, processing, storage, and marketing of agriculture products to shipment and after-sales services. In developing e-commerce for agriculture products, the provinces adopted a three-tier service facility system, which consists of county service centers, township service stations, and village service sites. Each service facility performs a different function. Villagers and farmers can sell their agriculture products, purchase goods online, make online payments, and pick up purchased goods at the service facilities without leaving their respective villages (ADB 2018).

E-commerce can potentially be more inclusive of underrepresented groups such as women, small businesses, and rural entrepreneurs. In China, for instance, Alibaba alone is reported to have created 30 million job opportunities, most notably among young people, in rural communities, and for disadvantaged groups (Kathuria et al. 2020). Alibaba's Taobao Villages have gradually spread inland, and in several cases, migrants have returned home to work in e-commerce—seizing job opportunities that would not have been available in rural areas without online platforms. These developments offer hope that e-commerce can be a powerful instrument for rural vitalization and poverty reduction (World Bank 2019a).

However, the food e-commerce sector may take time to develop, especially in connecting with rural areas. Not all food e-commerce initiatives have been successful in Asia—businesses have been closed, sold, or scaled back to focus on a handful of cities or countries (for example, Indonesia and the Philippines) (Green 2018). In China, rural e-commerce has faced a general lack of professional internet-savvy personnel. Thus, farmers' cooperatives and enterprises usually resort to third-party e-commerce platforms, such as Alibaba, Jingdong, Suning, Taobao, and Tmall, in starting their online businesses. These platforms are observed to tender low profits and maintain prohibitive access conditions (for example, collection of a security deposit upon registration and high fees for marketing activities) (ADB 2018).

persistent and changing dietary challenge. Undernutrition causes illnesses ranging from anemia to cognitive impairment and stunting, hurting both individuals and economies. In Vietnam, vitamin and mineral deficiencies reduce gross domestic product by more than $544 million annually (World Bank, n.d.). Moreover, a multicountry study in Southeast Asia finds that both underweight and obese children had lower IQ (intelligence quotient) scores than healthy-weight children (World Bank 2018). In Indonesia, the current prevalence of underweight children, at 19.6 percent, is quite high for a country that achieves economic growth of more than 5 percent per year (IFPRI 2019).

Fortification

Fortification of food with vital nutrients is becoming a mainstream activity in developing East Asia. It plays an increasingly important role in addressing population health concerns, from micronutrient deficiency to the challenges of an aging population in Asia (Green 2018). Government responses to micronutrient deficiencies and food fortification efforts using staple foods and condiments have advanced in Asia since the 1990s (box E.2 in appendix E) (Green 2018). A few countries in East Asia have also embraced biofortification, which is a "natural" way to introduce higher levels of micronutrients to diets.

Biofortification

Biofortification, that is, development of nutritionally enhanced staple crops, has been pursued since 1993 to improve both crop productivity and consumer health. Biofortification increases the density of vitamins and minerals in a crop, through plant breeding or agronomic practices, so that regular consumption will generate measurable improvements in vitamin and mineral nutritional status. Today, biofortified crops have been released in more than 30 countries and are being tested and grown in more than 40 countries (HarvestPlus in table D.1 in appendix D). Crisis situations have offered an opportunity to distribute new technology to smallholders (for example, in Africa, vitamin A–rich sweet potato in Mozambique after floods). In countries where biofortified varieties of widely consumed staple crops have already been approved for release (for example, Indonesia, the Philippines), the propagation and distribution of biofortified varieties can be implemented to help mitigate the rise in mineral and vitamin deficiencies under COVID-19 conditions—at no extra cost (IFPRI 2020).

Reformulation of food, functional foods, and nutrigenetics

Reformulation of food, functional foods, and nutrigenetics are mostly upstream activities in East Asia, and to some extent globally. Reformulation refers to the removal of harmful substances or finding natural replacements for ingredients such as sugar and fat. Indonesia, Korea, Singapore, and Thailand are currently on the verge of exploring reformulation to develop more healthy foods. Functional foods are processed foods with ingredients that provide substantial health benefits in addition to meeting basic nutritional needs.[19] Functional foods with healthy properties are receiving increasing interest in aging China and Japan (Ecosperity 2018b; Kotilainen et al. 2006). Nutrigenetics is a relatively novel area of innovation—it identifies how genetic variations affect people's responses to nutrients, the ultimate objective being personalized nutrition (WEF 2018). Nutrigenetics is still in the early phases of development given that the impact of certain genes on absorbing, transporting, storing, or metabolizing nutrients is not yet sufficiently understood (for more details, see appendix E).

Alternative food sources

Interest in food innovations that can lower the environmental footprint of food while providing sufficient nutrients and calories is increasing. The COVID-19 pandemic has brought greater attention to existing trends toward identifying and developing safe, healthy, affordable, and sustainable protein sources.

Alternative proteins are derived from nonlivestock sources—such as insects, plants (legumes, Kernza), aquaculture, and cell cultures (lab-grown meat)—and offer promising alternatives to traditional proteins used in human, animal, fish, and pet consumption. Given that beef has almost 30 times the GHG emissions per calorie of vegetables such as lentils or lab-grown meat, this shift can have a potentially significant impact on overall GHG emissions and the environmental footprint (box 5.12). The insect protein market alone is estimated to be worth about $8 billion by 2030, up from less than $1 billion today (Financial Times 2020b).

BOX 5.12

Lab-grown meat and other protein alternatives

Interest in emerging innovations and alternative protein sources is growing. Lab-grown meat may meet consumer needs without the heavy resource costs of rearing livestock, including the deforestation required to create grazing land, the atmospheric damage from livestock emissions, and the threat of zoonoses. Although lab-grown meat has yet to reach mainstream consumers, progress in making it affordable and accessible shows its potential to complement existing food production methods. Efforts to create artificial meat and protein have attracted investment from future-looking billionaires such as Richard Branson and Bill Gates as well as from large food firms. Lab-grown meats were forecast to be accessible to mainstream consumers by 2020 (Green 2018). Globally, more than two dozen firms are testing lab-grown fish, beef, and chicken, hoping to break into an unproven segment of the alternative meat market, which Barclays estimates could be worth $140 billion by 2029 (Meat News 2020).

Asia is an active player in lab-grown meat. Japan is leading Asia in lab-grown meat and cellular agriculture research. Acknowledging technological progress in other regions and the benefits for domestic meat consumption, China's government recently signed a $300 million trade deal with Israel, a lab-grown meat innovator, enabling the purchase of lab-grown meat from the country's suppliers. Media reports indicate that Silicon Valley–based Hampton Creek has also been in talks to license lab-grown meat technology to South Asia (Green 2018). In late 2020, the company, Eat Just Inc., has gained regulatory approval in Singapore to produce and sell lab-grown chicken meat (Meat News 2020).

Other emerging (or old in a new format) food innovations include insects, which are consumed in powdered form and contain more protein and micronutrients per pound than beef, and Kernza, a perennial plant that produces grain for five years (as opposed to wheat's single year of production) and can be used in baking and beer production. It has numerous ecological benefits because of its deep roots, which provide drought resilience as well as deposit carbon into the soil and boost overall soil health. Even algae are being used to produce a new form of oil, which could potentially act as a substitute for palm oil (Ecosperity 2018a).

Alternative proteins are still in an early stage of adoption and understanding, and may come with ancillary implications that require a systems perspective: if they prove popular, they could negatively affect the livelihood of livestock farmers and the economies of countries dependent on livestock, highlighting the need to account for trade-offs and externalities associated with this demand shift. Moreover, the health implications of the novel processes and ingredients used in some of these products are not yet well understood (WEF 2018). Consumers are also likely to take time to adjust to alternative foods such as lab meat as well as other innovations like insect protein products. There are also barriers to implementing nutrition education in countries such as the Republic of Korea, including the lack of a systematic curriculum and the lack of continuing education for nutrition teachers (Asian Scientist 2017; Le Page 2017; Woo et al. 2015). A final challenge is the scaling up of lab-meat production given that laboratory facilities cannot easily move from prototypes to mass market output.

The transition to better nutrition and alternative sources of nutrition faces several challenges, however, including low-income consumers' ability to pay, lack of access to healthier food (for example, time limitations to prepare food), information failures (understanding what is healthy and what is not), and the significant behavioral shifts required. For instance, although insects require much less feed and water, are a good source of energy and micronutrients, and produce less emissions, there are still microbial risks, allergen concerns, and consumer acceptance challenges (Asian Scientist 2017; Le Page 2017; Woo et al. 2015). Government policies that favor production of cereals have also encouraged the consumption of cereal-based foods over foods such as fruit and vegetables with higher nutrition value, which are relatively more expensive.

THE ECONOMIC, ENVIRONMENTAL, HEALTH, AND SOCIAL FEASIBILITY OF INNOVATIONS

The potential of transformative innovations to address the multiple challenges facing the agri-food system varies with each innovation and each context. Table 5.1 summarizes the key areas of innovation along the food supply chain (production, food chain management, and food consumption and nutrition) and the "traditional" and transformative innovations and their potential to address productivity, environmental sustainability, health, and inclusivity of food systems.

Innovations fostering agricultural production have great potential to improve agri-food system sustainability and resilience to risks and external shocks. These innovations offer a "triple win," that is, nearly all can address productivity, sustainability, and health aspects. The health aspects of mechanization and agricultural machinery (other than precision agriculture) are less straightforward, however; they may contribute to improved health by way of reducing physiological constraints of labor. Collective action, for example, through farmer organizations, out-grower schemes, and water users associations,

TABLE 5.1 Transformative innovations in agriculture and their potential to address productivity, sustainability, health, and inclusion issues

INNOVATION AREAS	PRODUCTIVITY	ENVIRONMENTAL SUSTAINABILITY	HEALTH, SAFETY, AND NUTRITION	EQUITY AND ECONOMIC FEASIBILITY[a]	ROLE OF PUBLIC AND PRIVATE SECTORS
		Production			
Precision agriculture	High	High	High	Up-front cost, size of land lots, land rights, skills	Both, mostly private
ICT in extension, weather, insurance, logistics services	High	High	Low	Cost, ICT connectivity	Both
Urban farming (vertical farming, hydroponics, aeroponics, growth kits) and protected farming	High	High	Moderate	Up-front cost, knowledge	Both
Biotechnology, GM (production traits, shelf life, micronutrient content, microbiomes, biopesticides, feed additives)	High	High	High	Seed or feed additive cost	Both

continued

TABLE 5.1, *continued*

INNOVATION AREAS	PRODUCTIVITY	ENVIRONMENTAL SUSTAINABILITY	HEALTH, SAFETY, AND NUTRITION	EQUITY AND ECONOMIC FEASIBILITY[a]	ROLE OF PUBLIC AND PRIVATE SECTORS
		Production			
Gene editing and other NBTs (production traits, shelf-life)	High	High	High	Seed cost	Both
Biofortified crops	High	Moderate	High	High feasibility	Both, mostly public
Climate-smart agriculture practices	High	High	High	Up-front cost, knowledge, skills	Both
Other sustainable agriculture approaches (pests, soil fertility, water, land-use intensity, livestock production, organic farming)	High	High	High	Knowledge, skills	Both
Mechanization (farm tools and machinery, irrigation, agro-services, 3D printing of agricultural tools)	High	Moderate	Moderate	Up-front cost, size of land lots	Private
		Food chain			
Traceability (blockchain, antifraud tools, sensors)	Moderate	Moderate	High	Costs, farm or firm ability to participate	Both, increasingly private
Food safety (diagnostic tests, antibacterial proteins)	n.a.	n.a.	High	Cost to firms	Both, increasingly private
Food loss and waste (data systems, e-commerce)	Moderate	Moderate	Low	Farm or firm ability to participate	Mostly private
Food loss and waste (storage solutions, smart labels, right-sizing, biodegradable packages)	Moderate	Moderate	Moderate	Up-front cost, energy	Mostly private
Food waste—circular economy (minimal and alternative use of waste)	Moderate	Moderate	Moderate	Up-front cost	Both, increasingly private
		Food consumption and nutrition			
Fortification	n.a.	n.a.	High	High feasibility	Mostly private
Reformulation and "free-from" foods, functional foods	n.a.	n.a.	High	Cost and access	Both, increasingly private
Nutrigenetics	n.a.	n.a.	High	Cost and access	Both, increasingly private
Lab-grown meat and other new food sources	Moderate	High	High	Cost and access, livelihood	Both, increasingly private
		Other			
Organization and collective action–related innovations (extension, value chain, water users association)	High	High	Moderate	High feasibility	Both

Source: Original table for this publication.
Note: GM = genetically modified; ICT = information and communication technology; NBT = new breeding techniques; High = high potential to address the challenge; Moderate = some potential to address the challenge; Low = low potential to address the challenge; n.a. = not applicable.
a. The ability of smallholders to benefit from new practices and tools and the ability to afford new nutritious food require attention to credit, skills development, organization of farmers, and development of affordable systems and innovations that smallholder farmers and farmer groups can adopt.

is often an important element of effective adoption of many production-oriented innovations that are more suitable for scale economies. Capital requirements or land ownership and structure may, however, be prohibitive to smallholder adoption of precision agriculture, mechanization, and some urban farming and CSA practices. Land consolidation or collective action among many farms may alleviate such constraints in reaching sufficient economies of scale. Similarly, some innovations (for example, CSA, precision agriculture, organic farming) require a higher level of skills and knowledge, making their adoption more challenging in isolated or poorly serviced locations and among the less educated and aging farmers. The higher costs of improved seed (for example, through biotechnology or GE) or feed additives, however, have often been offset by reduced costs (of labor and pesticides) and higher yields and improved growth and health of livestock.

Innovations in the food chain have the potential to improve health outcomes (for example, by reducing foodborne disease), and through improved traceability, to reduce FLW and GHG emissions. They may also to some extent improve productivity (improved efficiency and reduced FLW). At present, costs may discourage firms, and benefits to farmers depend on their ability to join trading and e-commerce schemes. Low-cost and collective storage solutions, however, offer great potential to smallholders.

The value of food consumption- and nutrition-related innovations lies primarily in their health aspects. Plant- and insect-based sources of protein also offer commercial opportunities for smallholders and may result in environmental benefits (fewer livestock and GHG emissions). Lab-grown meat also promises commercial opportunities, environmental sustainability, and health outcomes. The COVID-19 pandemic has further increased consumer interest in alternative meats. However, at present, such technology is beyond the reach of most firms and farmers. Costs and access to products with enhanced consumption and nutrition characteristics may be prohibitive and limit uptake by less affluent members of society, although much effort goes toward reducing the cost to consumers of such alternative sources of protein.

READINESS OF DEVELOPING EAST ASIAN COUNTRIES TO EMBRACE TRANSFORMATIVE INNOVATIONS

This section discusses the developing East Asian countries' capacity to embrace specific transformative innovations. Additional details on country capacity are provided in appendix E.

Overall capacity for transformative innovations across East Asia

The transformative innovations vary by level of science, technology, and innovation (STI) capacity required for their development as well as the level of ease with which they can be adopted in the existing agri-food system context, that is, considering farmer-enterprise capacity, access to innovation and finance, infrastructure, and so on. Figure 5.2 presents a simplified and subjective ranking of a range of innovations and the associated STI capacity requirements and ease of adoption. For instance, easy-to-adopt innovations include fortification, varieties, and agrochemicals, whereas greater skills and effort are required to adopt many forms of precision agriculture, food safety applications, and urban farming

FIGURE 5.2

Technology adoption and science, technology, and innovation capacities needed to use or develop different types of agricultural innovations

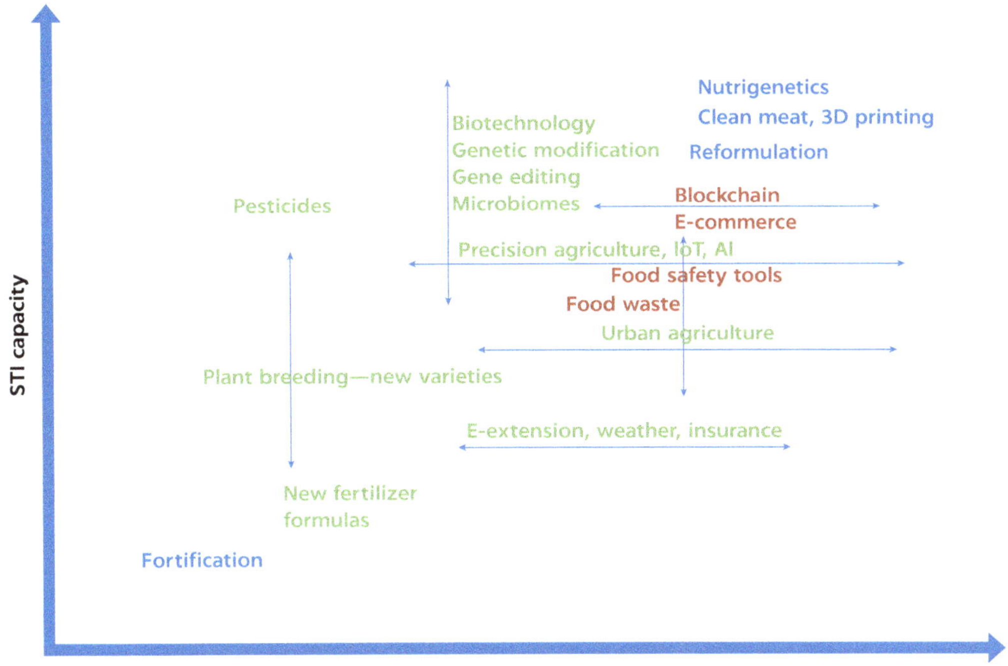

Source: Original figure for this publication, based on review of countries' innovation capabilities in agriculture.
Note: Green, red, and blue colored text refer to production, food chain, and food consumption/nutrition-oriented innovations, respectively. The arrows indicate an illustrative range of STI and diffusion/adoption capacities required to undertake the respective technology or innovation. 3D = three-dimensional; AI = artificial intelligence; IoT = Internet of Things; STI = science, technology, and innovation.

innovations. Biotechnology and many nutrition and digital approaches require high STI capacity whereas fortification, fertilizers, and traditional plant breeding require relatively less STI capacity.

The developing East Asian countries' innovation capacity as measured by their STI ability to generate and adopt a wide range of innovations varies (chapter 6). Figure 5.3 is a subjective assessment of East Asian countries' capacity to generate and adopt transformative innovations, drawing on the evidence about traditional and transformative innovations in the region and the countries' overall agricultural research and development and extension capacity. The overall frontrunners include the HICs of the region (Japan, Korea, and Singapore), closely followed by China, Malaysia, and Thailand. Although the countries have high STI capacity in many areas, some technologies and innovations are still beyond their reach (for example, lab-grown meat, nutrigenetics). All countries have capacity to generate simpler innovations such as fortification, fertilizers, and traditional plant breeding. Indonesia, the Philippines, and Vietnam also have the potential to catch up with China, Malaysia, and Thailand on, for example, STI in precision agriculture and urban farming. Although it may be more challenging for Cambodia, the Lao People's Democratic Republic, and Myanmar to catch up in STI capacity, they have the potential to continue generating and adopting traditional innovations and quickly catch up on adoption of less advanced and less costly e-services, urban agriculture, and precision agriculture.

FIGURE 5.3

Illustrative schema of technology adoption and science, technology, and innovation capacities in East Asian countries

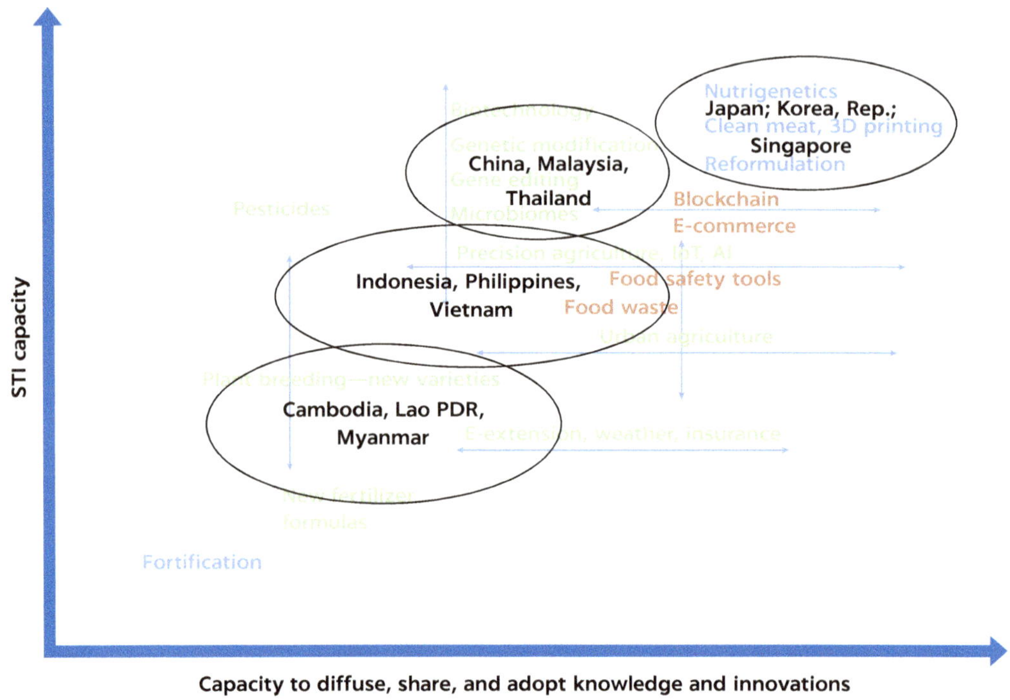

Source: Original figure for this publication, based on review of countries' innovation capabilities in agriculture.
Note: Green, red, and blue colored text refer to production, food chain, and food consumption/nutrition-oriented innovations, respectively. The arrows indicate an illustrative range of STI and diffusion/adoption capacities required to undertake the respective technology or innovation. 3D = three-dimensional; AI = artificial intelligence; IoT = Internet of Things; STI = science, technology, and innovation.

Innovation-specific capacity among the developing East Asian countries

Capacity in biotechnology largely correlates with economic development and country views on use of biotechnology, essentially GM and GE (table 5.2).[20] The region has significant experience in biotechnology, China being at the forefront on GE. China, Malaysia, and Thailand have the highest overall capacity to generate and adopt biotechnology innovations, but they have different views on the use of biotechnology (details in appendix E). Indonesia, the Philippines, and Vietnam continue improving their research and development and adoption capacity in biotechnology, mostly GM. Cambodia, Lao PDR, and Myanmar currently have low overall capacity in biotechnology but would be able to adopt, for example, crops with enhanced production traits.

Continued investment in biotechnology capacity is warranted in the region. Response to climate change and demands for productivity, food safety (for example, longer shelf life), and environmental sustainability (such as lab-grown meat, livestock breeds, feed) require high STI capacity. In addition to strengthening research capacity, biosafety regulatory issues (table 6.4 in chapter 6) warrant attention. Differences in regulations between countries have, however, created potential barriers to the use of GE.

TABLE 5.2 Country readiness to develop or adopt transformative innovations in crop, livestock, and aquaculture production

	CAPACITY FOR TRANSFORMATIVE INNOVATIONS IN AGRICULTURAL PRODUCTION[a]				
COUNTRY OR GROUP	BIOTECHNOLOGY, GM, GENE EDITING, MICROBIOMES	PRECISION AGRICULTURE AND OTHER IOT	ICT IN EXTENSION, WEATHER, FINANCE, LOGISTICS	URBAN FARMING	CSA AND OTHER SUSTAINABLE PRACTICES
Cambodia	Low	Low	Low-medium	Low-medium	Low-medium
China	Very high	High	High	High	High
Indonesia	Medium-high	Low-medium	Medium	Medium	Medium
Lao PDR	Low	Low	Low-medium	Low	Low-medium
Malaysia	High	High	High	High	High
Myanmar	Low	Low	Low	Low	Low
Philippines	Medium-high	Medium	Medium	Medium	Medium
Thailand	High	High	High	Medium-high	Medium-high
Vietnam	Medium-high	Medium	Medium	Medium	Medium-high
EAP Region (rest)	High	High	High	High	High
HICs	High	High	High	n.a.	High

Source: Original table for this publication.

Note: CSA = climate-smart agriculture; EAP = East Asia and Pacific; GM = genetically modified; HIC = high-income country; ICT = information and communication technology; IoT = internet of things; n.a. = not applicable.

a. Capacity for biotechnology category draws on the assessment in FAO (2019c) and the desk review. High capacity may refer to capacity to generate biotechnology but also a country's ability or desire to adopt biotechnology applications from elsewhere. For this reason, capacity of, for example, the Philippines and Vietnam is assessed at a higher level than expected. Although Myanmar has expressed interest in promoting, for example, GM and gene editing, it is currently ranked 34th place of 62 countries regarding the enabling environment for plant breeding, variety registration, and seed quality control in 2016 (World Bank 2017b).

Developing East Asian countries' capacity in precision agriculture varies, but all can adopt the least complex applications[21] (table 5.2). Overall, commercial adoption of precision tools has been slow. Although Malaysia tends to be the frontrunner among its peers in developing and adopting precision agriculture, China has also invested in precision agriculture and overall mechanization of agriculture (appendix E). Many countries have aging farming populations, suggesting growing demand for labor-saving innovations but also hurdles to adoption.

The challenge with precision agriculture will be to adapt agricultural technologies to the region's agro-ecological and small-farm realities, including aging farmers. Structural issues, such as limited land plots and rights, affordability, and inadequate infrastructure in rural areas, limit the overall market for and uptake of precision agriculture and e-services, driving the need for targeted interventions to expand rural access to ICT and mobile services. With focused research and investment (for example, farmer access to data, connectivity and automation, collective action for purchase of precision services, training and financial services to farmers), East Asia could lead the world in developing precision agriculture for smallholders. The role of the private sector is particularly prominent in development and commercialization of precision agriculture. Long-term development by the private sector is, however, constrained by lack of regulations for the use of the Internet of Things in agriculture, finance, and ICT connectivity.

Most East Asian countries have tried to optimize the outreach and coverage of extension with diverse e-extension services (table 5.2). However, the level of integration into national extension systems varies, and is rather limited in scope among the lower-middle-income countries (table 6.3 in chapter 6). China is the regional leader in promoting ICT in agriculture and has also integrated e-extension into its nationwide decentralized extension system. However, wider access to e-services in the region requires reliable ICT infrastructure, affordability of devices, capacity of providers and farmers, and public-private collaboration on content and delivery (for further details see appendix E and the section titled "Returns to innovation increased by reform of agricultural extension services and integration of e-extension" in chapter 6).

Urban farming is most sophisticated and widespread among the most urbanized countries of the region but is expected to take greater hold elsewhere as urbanization accelerates (table 5.2). Along with Malaysia, rapidly urbanizing China is ahead of its less affluent neighbors in promotion, research, and solutions in urban farming. The innovations and technical advice on urban farming are gradually taking hold in the less urbanized follower countries (appendix E). The common challenges to urban farming include regulatory issues (such as quality control), land availability, access to innovations and extension services, finance, and consumer acceptance. In addition to enabling innovation in technology and adoption (for example, incentives to the private sector), governments can use a range of levers to encourage urban agriculture, from streamlining regulations (safety, quality) to providing extension services and funding.

Most developing East Asian countries are making progress in developing or adopting CSA and other sustainable agriculture approaches (table 5.2). However, wide adoption of the practices has been met with challenges. Countries with strong research and development systems or functioning decentralized extension services,[22] supported by the private sector and international agricultural research centers, tend to be better off. A particular need is effort toward strengthening the technical capacity and coverage of the decentralized extension that is largely responsible for supporting farmers in CSA. In addition, private sector services (inputs, advisory services, buyers) could be enabled (with knowledge and incentives, including price incentives for certified "green products") to better reach out to farmers and support the adoption of CSA. Eventually, adoption of CSA and other sustainable innovations depends on farmers' capacity to adopt these innovations. More-educated farmers adopt new technologies early and get more profit out of them, an idea embraced by China (Professional Farmer) and Thailand (Smart Farmer) (appendix F).

Capacity in traceability and food safety innovations varies significantly among the developing East Asian countries[23] (table 5.3). Apart from China, Malaysia, and Thailand, most developing East Asian countries are still far from generating or adopting innovative tools in traceability and food safety. Currently, China leads (with blockchain at the state level) in East Asia, while Malaysia (Penang state) and Thailand are about to adopt blockchain for tracking agri-food chains. In Indonesia, the Philippines, and Vietnam, capacity is still limited, and it is nearly nonexistent in Cambodia, Lao PDR, and Myanmar. Vietnam, however, intends to invest in emerging technologies, including blockchain. Some Vietnamese start-ups have applied new technologies to the supply chain (Austrade 2019). E-commerce is showing an uptick as a response to COVID-19, and there is reason to believe that the challenges of economic viability and limited farmer participation may be solved.

TABLE 5.3 Country readiness to develop or adopt transformative innovations in food chain dynamics and food consumption and nutrition

	INNOVATIONS IN FOOD CHAIN DYNAMICS		INNOVATIONS IN FOOD CONSUMPTION AND NUTRITION		
COUNTRY OR GROUP	TRACEABILITY AND FOOD SAFETY[a]	FOOD LOSS AND WASTE	FORTIFICATION	REFORMULATION, FUNCTIONAL FOODS	NUTRIGENETICS, LAB-GROWN MEAT, AND NEW FOOD SOURCES
Cambodia	Low	Low	Medium	Low	Low
China	Medium-high	Medium	High	Medium-high	Medium
Indonesia	Low-medium	Low-medium	High	Low-medium	Low
Lao PDR	Low	Low	Medium	Low	Low
Malaysia	Medium-high	Medium	High	Medium-high	Medium
Myanmar	Low	Low	Low-medium	Low	Low
Philippines	Low-medium	Low	High	Low-medium	Low
Thailand	Medium-high	Low-medium	High	Medium	Medium
Vietnam	Low-medium	Low	High	Low	Low
EAP Region (rest)	Very high	Medium-high	High	—	Medium
HICs	Very high	High	High	High	Medium-high

Source: Original table for this publication.
Note: — = not available; EAP = East Asia and Pacific; HIC = high-income country.
a. Based on Jaffee et al. (2019), World Bank (2019b), and other documents on countries' ability and responsiveness to new innovative approaches in food safety.

The weak foundations and sizable gap between food safety capacity and requirements are especially challenging among the rapidly urbanizing countries of Indonesia, the Philippines, and Vietnam. The public sector can enhance traceability and food safety through improved communication, adequate regulatory frameworks (see the section titled "Importance of the enabling environment to innovation" in chapter 6), certification, new diagnostic tools, and training of high-skilled people. The private sector has been more active in pursuing actual innovations, and would benefit from training; clarity on regulations, for example, for blockchain technology use in agriculture; improved infrastructure; and public-private partnerships for investment in heavy applications. Ensuring traceability at the farm level will also require sensors that farmers can afford.

Capacity in FLW is still limited in most developing East Asian countries. Most innovation in FLW and the circular economy takes place in HICs. In East Asia, Japan, Korea, and Singapore, as well as China and Malaysia, are leading the way in pursuing FLW solutions. The stakeholders in Indonesia, the Philippines, and Vietnam have become conscious of FLW, but only limited adoption or innovation currently takes place. Although awareness and momentum for addressing FLW exists in most countries, it needs to be built into national policies and strategies (food security and safety, environment, climate change, urban planning). Regional approaches, for example, to logistics, may also pave the way for the lower-income neighbors to benefit. In addition, the private sector should be enabled to invest in diverse food marketing and FLW management solutions (for example, cold storage solutions to smallholders, recycling).

All East Asian countries have the capacity to fortify foods, but capacity for more sophisticated nutrition innovations, such as reformulation, functional foods, and nutrigenetics, is limited (table 5.3). Capacity, or perhaps propensity, for reformulation is mostly concentrated in the HICs of Korea and Singapore, and to some extent in Indonesia and Thailand (Green 2018). At the same time, aging China and Japan have shown interest in functional foods. The transition to better nutrition and alternative sources of nutrition would require improving consumer awareness and access to alternative foods. For instance, in the long term, food manufacturers are likely to focus on reformulation to meet demand for healthier alternatives.[24] However, becoming a leader in such novel approaches to nutrition and food would require focused investment in cross-disciplinary research and innovation, training and attracting chemists, life scientists, food scientists, marketers, nutrition experts, and agribusiness companies to this field. Yet this is an area with scant information, suggesting cross-disciplinary collaboration is limited or still under the radar.

NOTES

1. Slow adoption can be explained by fragmented rural markets, poor infrastructure, high regulatory burdens, and other factors that raise costs, while revenues are constrained by customers' limited ability and willingness to pay. Much of the start-up activity is in high-income countries across the world, indicating both the risk of unequal access to new solutions and the opportunities for scaling up in developing countries (WEF 2018).
2. These transformative innovations include e-services and sustainable agriculture practices for producers; vertical farming; interest in automation, artificial intelligence, precision agriculture to offset disruptions in labor supply, and drones to address disruptions in logistics; innovations for food packaging, traceability and accountability, e-commerce, and food waste solutions; and nutrition solutions (for example, products that enhance the immune system, and plant-based and alternative proteins).
3. E-services have the potential to reduce greenhouse gas emissions by 100 metric tons of carbon dioxide equivalent and freshwater withdrawals by 100 billion cubic meters (WEF 2018).
4. Crop insurance could provide farmers with lower risk and improved access to credit. In high-income countries, 1.99 percent of agricultural GDP is spent on agricultural insurance, but that share falls to 0.29 percent in upper-middle-income countries, 0.16 percent in lower-middle-income countries, and 0.01 percent in low-income countries (Cornell University, INSEAD, and WIPO 2017).
5. A meta-analysis of e-extension services suggests that providing agricultural information to farmers or extension agents in India, Kenya, Nigeria, and Uganda increased yields by 4 percent on average (Fabregas, Kremer, and Schilbach 2019).
6. A distributed ledger is a type of database, or system of records, that is shared, replicated, and synchronized among the members of a network. A blockchain is a type of distributed ledger that is composed of unchangeable, digitally recorded data in packages called blocks, where each block is then "chained" to the next block, using a cryptographic signature (FAO 2019a).
7. High scientific capacity is also required for lab-grown meat and 3D (three-dimensional) printing applications in manufacturing of food and machinery.
8. Traits include abiotic stress tolerance, cooking quality, virus or bacterial resistance, fiber content, modified oil composition, nutrition quality, pollination control, and yield (based on 2016 data from FAO 2019c).
9. Microbiomes have the potential to generate up to $100 billion in additional farmer income, increase production by up to 250 million tons, and reduce GHG emissions by up to 30 metric tons of carbon dioxide equivalent (WEF 2018).
10. About 266 million households in developing countries (more than 65 percent from Asia) are engaged in urban farming, and a quarter earn an income from it (Hamilton et al. 2014). With a forecast of nearly 90 percent urbanization by 2050, urban farming could contribute to nutrition (UN 2014).

11. For instance, Qualvis is using fully compostable packaging for its vegan chocolate, and Fazer and Sulapac have collaborated to create a compostable, microplastics-free box for Fazer's handmade pralines. Recently, Air New Zealand conducted a trial with edible coffee cups both on flights and on the ground, with the aim of reducing its use of single-use coffee cups.
12. A plastic material is defined as a bioplastic if it is bio-based, is biodegradable, or features both properties (European Bioplastics 2018).
13. Edible food thrown out by restaurants in China is equivalent to nearly 10 percent of the country's annual crop production.
14. Initiatives by the African Union, the United Nations, the Food and Agriculture Organization of the United Nations, the World Food Programme, and the World Bank are discussed at http://datatopics.worldbank.org/what-a-waste/global_food_loss_and_waste.html.
15. In 2017, China's overall e-commerce market ranked first in the world in both sales and growth (ADB 2018).
16. "Local Food Systems and COVID-19: A Look into China's Responses" (http://www.fao.org/in-action/food-for-cities-programme/news/detail/en/c/1270350/).
17. Ibid.
18. Indonesia's processed meat and poultry markets grew at an annualized rate of 27 percent between 2011 and 2015 (Ecosperity 2018a).
19. Functional foods include food products containing probiotics, prebiotics, plant stanols, and folic acid. Some foods such as salmon and rye-based foods are products that naturally contain functional properties (Agriculture and Agri-Food Canada 2014; Kotilainen et al. 2006). Market research estimates that the market value for functional food was $44 billion in 2015 (Ecosperity 2018b).
20. The biotechnology (biofertilizers, biopesticides, tissue culture, GM, and GE) capacity assessment is based on the method applied by FAO (2019b) and the desk review (attention to awareness and vision, commitment, action, and skills for each innovation area).
21. The application of 3D printing to the creation of agricultural tools and machinery is already a growing industry in the United States, and there is evidence of 3D-printed farm technology in low-income contexts such as Myanmar (Harimoto 2016).
22. For instance, some areas of China, Vietnam, plantation crop sectors in Malaysia, Thailand, and Indonesia have well-functioning extension services.
23. This assessment draws on the Food Safety Initiative (World Bank 2019b) and the desk review (including readiness to adopt).
24. About 1 percent of the population suffers from celiac disease (an immune reaction to eating gluten) in Indonesia (Green 2018).

REFERENCES

ADB (Asian Development Bank). 2018. *Internet Plus Agriculture: A New Engine for Rural Economic Growth in the People's Republic of China.* Metro Manila, Philippines: ADB.

Agriculture and Agri-Food Canada. 2014. "Fortified/Functional Foods in China." http://www.agr.gc.ca/eng/industry-markets-and-trade/international-agri-food-market-intelligence/asia/market-intelligence/fortified-functional-foods-in-china/?id=1414775892199.

Arizona State University. 2017. "ASU Students Work with Himalayan Farmers on Irrigation Solutions." ASU Now, July 14, 2017. https://asunow.asu.edu/20170714-global-engagement-asu-students-himalayan-farmers-irrigation-solutions.

Asian Scientist. 2017. "CRISPR Used to Edit Genes in Human Embryos." https://www.asianscientist.com/2017/08/topnews/correction-gene-human-embryos/.

Austrade. 2019. "Vietnam's Innovation Ecosystem." Commonwealth of Australia, Austrade, Hanoi and Ho Chi Minh City.

Bedford, M. R. 2018. "The Evolution and Application of Enzymes in the Animal Feed Industry: The Role of Data Interpretation." *British Poultry Science* 59 (5): 486–93. https://doi.org/10.1080/00071668.2018.1484074.

Bloomberg. 2017. "'Plant Factories' Churn Out Clean Food in China's Dirty Cities." Bloomberg News, May 25, 2017. https://www.bloomberg.com/news/features/2017-05-25/-plant-factories-churn-out-clean-food-in-china-s-dirty-cities.

Burrellon, Heather. 2019. "Top Three Food and Beverage Packaging Trends for 2020." FoodBevMedia, December 24, 2019. https://www.foodbev.com/news/top-3-food-and-beverage-packaging-trends-for-2020/.

Cai, Jinyang, Ruifa Hu, Jikun Huang, and Xiaobing Wang. 2017. "Innovations in Genetically Modified Agricultural Technologies in China's Public Sector: Successes and Challenges." *China Agricultural Economic Review* 9 (2): 317–30. https://doi.org/10.1108/CAER-10-2016-0170.

Callaway, Ewen. 2018. "CRISPR Plants Now Subject to Tough GM Laws in European Union. Top Court's Ruling Threatens Research on Gene-Edited Crops in the Bloc." *Nature*, July 25, 2018. https://www.nature.com/articles/d41586-018-05814-6.

Cassou, E., S. M. Jaffee, and J. Ru. 2017. *The Challenge of Agricultural Pollution: Evidence from China, Vietnam, and the Philippines*. Directions in Development. Washington, DC: World Bank.

Cohen, J. 2019. "To Feed Its 1.4 Billion, China Bets Big on Genome Editing of Crops." *Science*, July 29, 2019. https://www.sciencemag.org/news/2019/07/feed-its-14-billion-china-bets-big-genome-editing-crops.

Cornell University, INSEAD, and WIPO (World Intellectual Property Organization). 2017. *The Global Innovation Index 2017: Innovation Feeding the World,* 10th ed. Ithaca: Cornell University; Fontainebleau: INSEAD; and Geneva: WIPO. https://www.insead.edu/sites/default/files/assets/dept/globalindices/docs/GII-2017-report.pdf.

Economic Consulting Associates Limited. 2020. "African Solar Designs. Market Assessment Study: Stand-alone Solar for Productive Uses in Uganda." Draft Final Report, February 13, 2020. Submitted to World Bank by Economic Consulting Associates.

Economist. 2020. "Using Artificial Intelligence, Agricultural Robots Are on the Rise." February 6, 2020. https://www.economist.com/science-and-technology/2020/02/06/using-artificial-intelligence-agricultural-robots-are-on-the-rise.

Ecosperity. 2018a. "Feeding Urban Asia: New Approaches for Providing Safe, Nutritious, and Affordable Food." Ecosperity, Singapore.

Ecosperity. 2018b. "Prevention Is the Best Medicine: Rethinking Healthcare in Asia." Ecosperity, Singapore.

EMF (Ellen MacArthur Foundation). n.d. "What Is a Circular Economy? A Framework for an Economy That Is Restorative and Regenerative by Design." https://www.ellenmacarthurfoundation.org/circular-economy/concept.

Espi, E., A. Salmeron, A. Fontecha, Y. García, and A. Real. 2006. "Plastic Films for Agricultural Applications." *Journal of Plastic Film and Sheeting* 22 (2): 85–102.

European Bioplastics. 2018. "What Are Bioplastics? Fact Sheet." European Bioplastics, Berlin. https://docs.european-bioplastics.org/publications/fs/EuBP_FS_What_are_bioplastics.pdf.

Fabregas, R., M. Kremer, and F. Schilbach. 2019. "Realizing the Potential of Digital Development: The Case of Agricultural Advice." *Science* 366 (6471). doi:10.1126/science.aay3038.

FAO (Food and Agriculture Organization of the United Nations). 2016. "Probiotics in Animal Nutrition: Production, Impact and Regulation." FAO Animal Production and Health Paper 179, FAO, Rome. http://www.fao.org/3/a-i5933e.pdf.

FAO (Food and Agriculture Organization of the United Nations). 2017. "Save Food: Global Initiative on Food Loss and Waste Reduction." FAO, Rome. http://www.fao.org/save-food/resources/keyfindings/en/.

FAO (Food and Agriculture Organization of the United Nations). 2019a. *E-Agriculture in Action. Blockchain for Agriculture. Opportunities and Challenges*. Bangkok: Food and Agriculture Organization of the United Nations and International Telecommunication Union.

FAO (Food and Agriculture Organization of the United Nations). 2019b. *Five Practical Actions towards Low-Carbon Livestock*. Rome: FAO.

FAO (Food and Agriculture Organization of the United Nations). 2019c. "The Status of Application, Capacities and the Enabling Environment for Agricultural Biotechnologies in the Asia-Pacific Region." Regional Background Study. FAO, Rome.

Feed the Future. 2017. "Digital Soil Test Kit." Presented at the Global Learning and Evidence Exchange, Market Systems, Bangkok, Thailand, March 21–24. https://agrilinks.org/sites/default/files/Asia%20Markets%20GLEE%20-%20Session%2015%20-%20Innovation%20Stations%20-%20 Ehsanul%20Karim.pdf.

Financial Times. 2020a. "Vertical Farming: Hope or Hype?" *Financial Times,* October 31, 2020. https://www.ft.com/content/0e3aafca-2170-4552-9ade-68177784446e.

Financial Times. 2020b. "Regenerative Nature." *Financial Times,* Dec 17, 2020. https://rethink.ft.com/articles/insect-proteins-big-push/.

Fuglie, K., Gautam, M., Goyal, A., and Faloney, W. F. 2020. *Harvesting Prosperity: Technology and Productivity Growth in Agriculture.* Washington, DC: World Bank.

Giovannucci, D., S. Scherr, D. Nierenberg, C. Hebebrand, J. Shapiro, J. Milder, and K. Wheeler. 2012. *Food and Agriculture: The Future of Sustainability.* New York: United Nations Department of Economic and Social Affairs. https://sustainabledevelopment.un.org/content/ documents/1443sd21brief.pdf.

Goedde, Lutz, Maya Horii, and Sunil Sanghvi. 2015. "Pursuing the Global Opportunity in Food and Agribusiness." McKinsey and Company. https://www.mckinsey.com/industries/chemicals/our-insights/pursuing-the-global-opportunity-in-food-and-agribusiness.

Green, Adam. 2018. "Food 4.0: The Future of Food Innovation in Asia." Economist Intelligence Unit Limited.

Hamilton, A. J., K. Burry, H.-F. Mok, S. F. Barker, J. R. Grove, and V. G. Williamson. 2014. "Give Peas a Chance? Urban Agriculture in Developing Countries: A Review." *Agronomy for Sustainable Development* 34: 45–73. https://link.springer.com/article/10.1007/s13593-013-0155-8.

Harimoto, T. 2016. "Farmers in Myanmar Are 3D Printing Their Own Tools." Makerbot, July 28, 2016. https://www.makerbot.com/media-center/2016/09/16/myanmar-farmers-3d-print-their-own-tools.

HelloTractor. 2019. "Company Brief. Growing Profitable Tractor Contracting and Farmer Yields. Quarter 2019." https://www.futurefarming.events/wp-content/uploads/2019/11/Priscilla-Asonibare-Hello-Tractor.pdf.

IFPRI (International Food Policy Research Institute). 2019. "Policies to Support Investment Requirements of Indonesia's Food and Agriculture Development during 2020–2045." Manila: IFPRI, Kementerian PPN/Bappenas, and Asian Development Bank.

IFPRI (International Food Policy Research Institute). 2020. "Getting Biofortified Food on Everyone's Plate: HarvestPlus 2019 Annual Report." IFPRI, Washington, DC. https://www.ifpri.org/publication/getting-biofortified-food-everyones-plate-harvestplus-2019-annual-report.

ISAAA (International Service for the Acquisition of Agri-biotech Applications). 2019. Global Status of Commercialized Biotech/GM Crops: 2019. ISAAA Brief 55. https://www.isaaa.org/resources/publications/briefs/55/.

Jaffee, S., S. Henson, L. Unnevehr, D. Grace, and E. Cassou. 2019. "The Safe Food Imperative: Accelerating Progress in Low- and Middle-Income Countries." World Bank, Washington, DC.

Jambeck, J. R., R. Geyer, C. Wilcox, T. R. Siegler, M. Perryman, A. Andrady, R. Narayan, and K. L. Law. 2015. "Plastic Waste Inputs from Land into the Ocean." *Science* 347 (6223).

Kang, Y., Y-C. A. Chang, H.-S. Choi, and M. Gu. 2013. "Current and Future Status of Protected Cultivation Techniques in Asia." SHS Acta Horticulturae 987: International Symposium on High Tunnel Horticultural Crop Production. https://www.actahort.org/books/987/987_3.htm.

Karplus, V. J., and X. W. Deng. 2006. "Agricultural Biotechnology in China: Past, Present and Future." In *Biotechnology and Sustainable Agriculture 2006 and Beyond, Proceedings of the 11th IAPTC&B Congress, August 13–18, 2006 Beijing, China,* edited by X. Zhihong, L. Jiayang, X. Yongbiao, and W. Yang, 81–86. Berlin: Springer.

Kathuria, S., A. Grover, V. M. E. Perego, A. Mattoo, and P. Banerjee. 2020. *Unleashing E-Commerce for South Asian Integration.* International Development in Focus. Washington, DC: World Bank. doi:10.1596/978-1-4648-1519-5.

Keating, K. 2019. "What the Food & Beverage Packaging Innovation Report Says About Packaging Innovation." *"We Listen" Blog*, October 30, 2019. https://www.pkgbranding.com/blog/what-the-food-beverage-packaging-innovation-report-says-about-packaging-innovation.

Kemira. 2020. "Future of Food Packaging: Four Scenarios for 2030." https://www.kemira.com/insights/future-food-packaging-four-scenarios-2030/.

Korhonen, J., A. Koskivaara, and A. Toppinen. n.d. "Future Pathways of Fiber-Based Packaging Sector." University of Helsinki. https://www.helsinki.fi/sites/default/files/atoms/files/korhonen_cegebe_fiber_based_packaging.pdf.

Kotilainen, L., R. Rajalahti, C. Ragasa, and E. Pehu. 2006. "Health Enhancing Foods: Opportunities for Strengthening the Sector in Developing Countries." Agriculture and Rural Development Discussion Paper 30, World Bank, Washington, DC.

Lamont, W. J. 2009. "Overview of the Use of High Tunnels Worldwide." *HortTechnology* 19 (1): 25–29.

Ledger Insights. 2020. 'Blockchain Agri Commodity Firm Cerealia Goes Live." *Ledger Insights*, November 10, 2020. https://www.ledgerinsights.com/blockchain-agri-commodity-firm-cerealia-goes-live/.

Le Page, M. 2017. "Boom in Human Gene Editing as 20 CRISPR Trials Gear Up." *NewScientist*, May 30, 2017. https://www.newscientist.com/article/2133095-boom-in-human-gene-editing-as-20-crispr-trials-gear-up/.

Liu, E., W. He, and C. Yan. 2014. "'White Revolution' to 'White Pollution'—Agricultural Plastic Film Mulch in China." *Environmental Research Letters* 9 (9): 091001.

Liu, Y., C. D. Espinosa, J. J. Abelilla, G. A. Casas, L. V. Lagos, S. A. Lee, W. B. Kwon, J. K. Mathai, D. Navarro, N. W. Jaworski, and H. S. Stein. 2018. Non-Antibiotic Feed Additives in Diets for Pigs: A Review." *Animal Nutrition* 4 (2): 113–25. https://doi.org/10.1016/j.aninu.2018.01.007.

Meat News. 2020. "Lab-Grown Chicken Meat Approved in Singapore." *Meat News*, December 22, 2020. https://www.poultryworld.net/Meat/Articles/2020/12/Lab-grown-chicken-meat-approved-in-Singapore-685811E/.

Meticulous Research. 2020. "Feed Additives Market—Global Opportunity Analysis and Industry Forecast (2019–2027)." Wakad, Pimpri-Chinchwad, Meticulous Research. https://www.meticulousresearch.com/product/feed-additives-market-5082/.

Mitchell, M. 2019. "Innovation with Livestock Feed: The Future of Animal Source Foods." *Agrilinks* (blog), October 2, 2019. https://www.agrilinks.org/post/innovation-livestock-feed-future-animal-source-foods.

Mondal, P., and M. Basu. 2009. "Adoption of Precision Agriculture Technologies in India and in Some Developing Countries: Scope, Present Status and Strategies." *Progress in Natural Science* 19 (6): 659–66. http://www.sciencedirect.com/science/article/pii/S1002007109000173.

Moreno-Peñaranda, Raquel. 2011. "Japan's Urban Agriculture: Cultivating Sustainability and Well-being." United Nations University, Tokyo.

Nhan, D. K., L. T. Duong, V. S. Nguyen, and M. Verdegem. 2005. "Development of 'VAC' Integrated Farming Systems in the Mekong Delta, Vietnam: A View of a System and a Participatory Approach." In *Development of Integrated Agriculture Farming Systems in the Mekong Delta*, edited by R. Yamada, 101–25. Ho Chi Minh City: Tre Publishing House. https://www.researchgate.net/publication/230594787_Development_of_VAC_Integrated_Farming_Systems_in_the_Mekong_Delta_Vietnam_-A_View_of_a_System_and_a_Participatory_Approach.

Ortolani, G., and P. Di Bella. 2015. "Simple Agricultural Innovation to Empower Farmers." SciDevNet. http://www.scidev.net/global/agriculture/multimedia/agricultural-innovation-empower-farmers-technology.html.

PwC and FIA (Food Industry Asia). 2020. "Maintaining Food Resilience in a Time of Uncertainty: Understanding the Importance of Food Value Chains in ASEAN and How to Ensure Their

Resilience during the COVID-19 Crisis." Price Waterhouse Coopers, Commissioned by Food Industry Asia.

RSIS Centre for Non-Traditional Security (NTS) Studies. 2014. *International Conference on Asian Food Security (ICAFS) 2014: Towards Asia 2025*. Singapore: Nanyang Technological University.

RUAF Foundation. 2018. "Urban Agriculture: What and Why?" http://www.ruaf.org/urban-agriculture-what-and-why.

Schuttelaar and Partners. 2019. "Report on Use of Plastics in Agriculture." Wageningen, the Netherlands, May 28, 2019.

Sim, Walter. 2018. "Goats, Pigs and Veggies Crop Up in Urban Tokyo." *Straits Times*, February 24, 2018. http://www.straitstimes.com/asia/east-asia/goats-pigs-and-veggies-crop-up-in-urban-tokyo.

Singh, Bryna. 2016. "Vertical Farms on the Rise in Land Scarce Singapore." *Straits Times*, July 10, 2018. http://www.straitstimes.com/lifestyle/vertical-farms-on-the-rise-in-land-scarce-singapore.

Smart Agriculture Analytics. 2015. "China's Indoor Agriculture Industry."

Teng, P. 2017. "Knowledge Intensive Agriculture: The New Disruptor in World Food?" RSIS Commentary 124, Nanyang Technological University. http:// www.rsis.edu.sg/wp-content/uploads/2017/06/CO17124.pdf.

Toop, T. A., S. Ward, T. Oldfield, M. Hull, M. E. Kirby, and M. K. Theodorou. 2017. "Agrocycle—Developing a Circular Economy in Agriculture." *Energy Procedia* 123 (2017): 76–80.

UN (United Nations). 2014. *World Urbanization Prospects: 2014 Revision*. New York: United Nations. https://esa.un.org/unpd/wup/publications/files/wup2014-highlights.pdf.

UNEP (United Nations Environment Programme). 2018. *Single-Use Plastics: A Roadmap for Sustainability*. Nairobi: UNEP. https://wedocs.unep.org/bitstream/handle/20.500.11822/25496/singleUsePlastic_sustainability.pdf?sequence=1&isAllowed=y.

UNSDSN (United Nations Sustainable Development Solutions Network). 2013. *Solutions for Sustainable Agriculture and Food Systems: Technical Report for the Post-2015 Development Agenda*. Paris: UNSDSN. http://unsdsn.org/wp-content/uploads/2014/02/130919-TG07-Agriculture-Report-WEB.pdf.

Vox, G., R. V. Loisi, I. Blanco, G. S. Mugnozza, and E. Schettini. 2016. "Mapping of Agriculture Plastic Waste." *Agriculture and Agricultural Science Procedia* 8: 583–91.

WEF (World Economic Forum). 2018. *Innovation with a Purpose: The Role of Technology Innovation in Accelerating Food Systems Transformation. System Initiative on Shaping the Future of Food Security and Agriculture*. Geneva: World Economic Forum together with McKinsey & Company.

Woo, J. W., J. Kim, S. Kwon, C. Corvalan, S. Cho, H. Kim, and J.-S. Kim. 2015. "DNA-Free Genome Editing in Plants with Preassembled CRISPR-Cas9 Ribonucleoproteins." *Nature Biotechnology* 33: 1162–64. https://www.nature.com/articles/nbt.3389.

World Bank. n.d. "Nutrition at a Glance: Vietnam." World Bank, Washington, DC. http://siteresources.worldbank.org/ NUTRITION/Resources/281846-1271963823772/Vietnam.pdf.

World Bank. 2010. *People, Pathogens, and Our Planet. Volume 1: Towards a One Health Approach for Controlling Zoonotic Diseases*. Washington, DC: World Bank.

World Bank. 2012. *Agricultural Innovation Systems. Investment Sourcebook*. Washington, DC: World Bank.

World Bank. 2017a. *ICT in Agriculture: Connecting Smallholders to Knowledge, Networks, and Institutions*. Washington, DC: World Bank. https://openknowledge.worldbank.org/handle/10986/27526.

World Bank. 2017b. *Myanmar's Public Expenditure Review: Fiscal Space for Economic Growth*. Yangon: World Bank Myanmar.

World Bank. 2018. "The Human Capital Project." In *World Development Report 2019: The Changing Nature of Work*. Washington, DC: World Bank. doi:10.1596/978-1-4648-1328-3.

World Bank. 2019a. *E-commerce Development. Experience from China.* Washington, DC: World Bank Group and Alibaba Group.

World Bank. 2019b. "*Managing Headwinds. East Asia and Pacific Economic Update.*" Washington, DC: World Bank. doi:10.1596/978-1-4648-1412-9.

World Bank. 2019c. "Marine Plastic Debris. A Regional Engagement Framework for Action for the World Bank East Asia and Pacific Region." World Bank, Washington, DC.

World Bank. 2020. "Marine Plastics." Separate briefs for Cambodia, Indonesia, Lao PDR, Malaysia, Philippines, Vietnam, Thailand. Unpublished, World Bank, Washington, DC.

Zhou, W. 2013. "Food Waste and Recycling in China: A Growing Trend?" Worldwatch Institute, Washington, DC. http://www.worldwatch.org/food-wasteand-recycling-china-growing-trend-1.

6 Enhancing Institutional Capacity for Innovation

INTRODUCTION

The evolution of agricultural innovation systems (AIS) in East Asian countries is typically aligned with the development of each country's agri-food sector and overall economy (figure 3.2). Countries can be divided roughly into three groups regarding their chosen strategy to foster productivity and sustainability and their overall institutional readiness to generate and adopt innovations. The groups are as follows.

The upper-middle-income countries (upper MICs) of China, Malaysia, and Thailand are more urbanized, and their agricultural share of gross domestic product (GDP) has decreased (table 1.1). They have greater research and development (R&D) spending (apart from Thailand), emphasize innovation and technology in their strategies,[1] and benefit from relatively strong private sector–led innovation activity, with actors ranging from large multinationals in input and food sectors to smaller start-ups and entrepreneurs in e-services, input supply, and trade. This private sector activity is also reflected in foreign direct investment (FDI) to the agriculture sector. All three countries have net FDI outflows (table 1.2). Since acceding to the World Trade Organization in 2001, China has gradually shifted from being a net exporter of food to a net importer of food (globally, the largest importer since 2014). China's productivity challenge is also partly addressed through significant net FDI to agriculture.[2]

The mid-range MICs of Indonesia, the Philippines, and Vietnam are urbanizing; but their agriculture sectors are less developed. Agriculture's share of the economy has also declined, but the sector is less sophisticated (less value added, lower private sector activity). Countries' visions and strategies are not always aligned with investment in agricultural R&D and innovation. For instance, both Indonesia and Vietnam emphasize innovation-driven sector development and growth; however, the gap between the vision and reality is still wide. Private sector activity, although important and increasing, particularly for plantation crops, is also more limited in volume and nature.

The lower MICs of Cambodia, the Lao People's Democratic Republic, and Myanmar are predominantly rural; and the agriculture sector plays an important role in their economies. They lag behind their neighbors in vision and strategy as well as the reality on the ground. Sector activities rely significantly on donor funding. Private sector activity is limited and concentrated in commercial crops and agrochemicals.

As stated in chapter 1, this AIS assessment is a desk review and is limited by data availability and the concept of AIS used in any given report. The review draws on existing literature but does not extensively repeat the findings in the consulted reports. This chapter addresses a few elements of AIS more thoroughly and mostly descriptively. R&D, extension, and private sector activity in all nine countries are addressed;[3] the other elements of AIS primarily center on China, Indonesia, Malaysia, the Philippines, and Vietnam.

MISALIGNMENT BETWEEN SECTORWIDE VISION AND INNOVATION CAPACITY, INVESTMENT, AND GOVERNANCE

East Asia's agricultural performance over the past 50–60 years reflects its inclination toward innovation, its willingness to embrace new technologies, and its capacity for transformation. The countries, however, vary significantly in their vision, capacity, and readiness to capitalize on the new opportunities that innovations in production, the agri-food chain, and nutrition offer (see the section titled "Readiness of developing East Asian countries to embrace transformative innovations" in chapter 5). Although investment in innovation—including skills, knowledge, and the wider enabling environment—remains essential, the direction of innovation activity and its alignment with agri-food systems' vulnerabilities and new societal functions need to be reassessed.

The country vision should define the sector strategy and investments in the agri-food system, including investment in agricultural R&D and innovation (table 6.1). Many countries (for example, China, Indonesia, Malaysia, Thailand, and Vietnam) have adopted new "policy frameworks" toward sustainability (box 6.1), and some have implemented incentives programs for green

TABLE 6.1 **Country vision for agri-food sector development**

COUNTRY	MAIN VISION
Cambodia	Productivity, diversification, and commercialization without footprint
China	Productive, safe, and ecological agriculture Innovation the most important engine to promote growth
Indonesia	Sustainability but maintaining productivity Innovation-driven economy
Lao PDR	Productivity without footprint, integration of climate change in programs
Malaysia	Maintaining productivity and improve sustainability and inclusiveness Innovation-based growth and production processes
Myanmar	Expecting to adopt an integrated approach, productivity without footprint
Philippines	Productivity without environmental footprint
Thailand	Increasing productivity and income with reduced footprint and improved food safety Value-based economy founded on creativity, innovation, and intellect
Vietnam	Sustainability but maintain productivity Innovation a key engine for transformation

Source: Original table for this publication. Based on country overall science, technology, and innovation strategy reports; agriculture strategy reports; or both.

BOX 6.1

Vietnam's new policy framework toward more sustainable and market-led agricultural development

The past reforms and agricultural growth in Vietnam have resulted in new challenges. Agricultural policies have been primarily geared toward expanding output to meet food security, economic growth, and trade targets. This direction has encouraged land expansion and intensification and more intensive use of inputs and other means to raise productivity. The ubiquity of the small farm has led to a fragmented production structure, and to high transaction costs (lack of collective action) both in provision of technical advice on sustainable practices and in monitoring farmer compliance with regulatory or private standards. However, gradual consolidation of commercial agriculture is increasingly responsible for aquaculture, livestock, and rice production (World Bank 2016).

A new vision and plan have been developed to address challenges with diversification, sustainability, quality, and value added of products, technological and institutional innovation, and climate change. The plan places greater emphasis on sustainable development, market-led and consumer-driven agriculture, and a shifting role for the government from being the primary investor or service provider to being the facilitator of investments and services provided by the private sector, community organizations, research institutions, commercial banks, and others (World Bank 2015). To attract more investment to the sector, the government has also issued a series of policies to offer incentives to both foreign and domestic investors in the form of financial support (tax exemption, preferential credit, trade promotion, and other policies) to support land access and farming contracts or to reduce postharvest losses. Foreign direct investment in agriculture has remained limited (1 percent of the total foreign direct investment registered) despite the policies set in place (Diem and Thuy 2019).

agriculture, training programs for extension agents and farmers, and research programs on diverse aspects of sustainability (World Bank 2018a). Relatively more food insecure countries, such as Cambodia, Lao PDR, Myanmar, and the Philippines, largely emphasize productivity increases but with a limited environmental footprint. This overall change toward sustainability needs to continue and deepen to also encompass zoonoses management and food safety if the countries' agri-food systems are to cope with the challenges and respond to new opportunities.

Knowledge intensity is a cross-cutting theme in agri-food system development in developing East Asia. Agriculture 4.0, smart agriculture, high-tech agriculture, and an innovation-driven economy are featured in countries' policy and strategy papers targeting productivity, value added, sustainability, and safety of agriculture.[4] Knowledge intensity has become an overarching element of innovation, whether innovations concern adoption of new agronomic practices or rely more on scientific endeavors. For instance, adoption of simple, environmentally friendly agronomic practices (such as most climate-smart agriculture) requires more than basic skills in literacy. The application of these practices tends to be fairly knowledge intensive, putting strains on the capacity of extension services and farmers alike. Use of precision agriculture tools and even e-services also requires more from the service providers and farmers. Use of biotechnology relies on high science, technology, and innovation (STI) skills. Knowledge accumulation also plays a crucial role in AIS, as illustrated in the AIS framework (figure 3.1) (Fuglie et al. 2020).

A successful transition to a knowledge- and innovation-driven agri-food system requires greater innovation capacity across the agri-food spectrum, including higher farmer education levels (see the section titled "Innovation capacity and skills for long-term sustainability" in this chapter). Better educated and trained farmers are more likely to make successful changes to farm practices and be more innovative (Foster and Rosenzweig 2010; Labarthe and Laurent 2009). Today, the farmers in many of the region's developing MICs (for example, China, Malaysia, Thailand, and Vietnam) are aging and have less education than younger generations, constraining the agri-food sector's ability to innovate and transform. Different contexts, however, require different approaches to farmer education. Extension, including e-extension services (along with improved basic education), may be in the best position to improve smallholders' knowledge of, for example, climate-smart agriculture approaches. However, in the transition-urbanizing contexts, greater emphasis on technical and vocational education and training-level approaches may be warranted. For instance, the share of farmers with basic and full training has gradually increased in the European Union.[5] In the Netherlands, most farmers already have either basic (64.6 percent in 2010) or full (6.6 percent) training in agriculture, reflecting the importance and sophistication of the agri-food sector (exports of more than $90 billion in 2018). In East Asia, the Republic of Korea, for instance, has increased the education level of its farmers (OECD 2018b).[6]

Moving up the innovation ladder requires that greater importance be placed on coordination and governance of AIS as part of economywide national innovation system (NIS) governance. Well-functioning innovation systems critically depend on how well governments can bring together and coordinate the activities of the various actors and stakeholders fundamental to advancing STI policies in various sectors of the economy. The integration of the AIS into overall governance of the NIS ensures better use of public funds and increased efficiency as well as attention to cross-cutting STI policies. As the maturity of AIS and its governance increases, so do the requirements for monitoring and evaluation to assess the impact of different policies and instruments. Box 6.2 discusses the evolution of AIS governance arrangements in many agriculture-based countries as a response to better understanding agricultural innovation and its importance for the sector and dependency on the wider economy and overall STI policies (see the section titled "Better resource use and innovation outcomes from stronger governance of AIS" in this chapter).

The jump from being a technology adopter or innovator country to being an inventor at the technological frontier can be challenging. Overall, investment remains the single most important determinant of exceptional long-term economic growth;[7] however, it should be channeled to new areas and backed by improvements in governance, education, and infrastructure (EBRD 2018). The AIS must also encompass actors and visionaries capable of such strategic decision-making and coordination. For instance, the Netherlands—historically, a prominent producer and exporter of horticultural products—has built on its strong base in horticulture and gradually shifted toward innovation and higher-value-added products with greater profit margins[8] (Dons and Bino 2008). Chile, an upper MIC, is undergoing such a transition process from being a successful technology adopter to being an inventor by strengthening its innovation capacity and AIS, including leadership among public and private sectors, and financing of diverse and strategically appropriate AIS (box 6.3). Thus, there is no one-size-fits-all policy mix for agri-food system transformation. However, developing East Asian countries with strong AIS capacity, national strategic leadership, capable governance, and available financing may be able to make such a transition.

BOX 6.2

Evolution of agricultural innovations systems governance

Research governance. The lack of appropriate coordination and governance for agricultural innovation systems (AIS) at the national level is a chronic problem for many countries. Historically, in the agriculture sector, the earliest attempts at coordinating AIS were centered on strengthening agricultural research coordination. Several countries have also established research governance bodies, for example, formal apex research councils to govern multi-institutional national agricultural research systems; but they tend to represent only a narrow range of AIS stakeholders, consisting primarily of ministerial representatives or researchers. They have often lacked a consistent, rigorous process for setting priorities. The current movement to improve the representativeness of these governance bodies and their mode of operation (for example, seeking to represent a wider range of stakeholders and regions, improving transparency, and using diverse prioritization tools) is encouraging.

Multidisciplinary national innovation system. The overall trend is toward strengthened research governance and multidisciplinary national innovation system governance (for example, the Thailand National Innovation Agency and, in the Republic of Korea, the Presidential Advisory Council for Education, Science, and Technology and the National Science and Technology Council), wherein agriculture is one sector among many. Despite their presence, national innovation councils and agricultural research councils rarely operate as true agricultural innovation organizations or councils, with a mandate to coordinate and prioritize investments in agricultural innovation at the highest level.

AIS governance. Some countries have made specific efforts in AIS-centric governance. The three commonly applied modes of coordinating innovation for agriculture at higher levels to date include the national agricultural innovation council or committee, competitive innovation or research funds, and coordination by theme or subsector. Typically, these efforts center on subsectoral governance and coordination—for instance, through commodity boards and subsector networks—rather than on national agriculture or rural innovation governance structures (such as Chile's Foundation for Agricultural Innovation and Australia's Rural Research and Development Council).

Sources: National bodies for coordination and governance of AIS in World Bank 2012.

BOX 6.3

Chile's long history of strategic choices on agri-food sector investments and innovation-led growth

From the 1960s onward, Chile implemented a dual policy aimed at both self-sufficiency and support to key commodities with comparative advantage (World Bank 2020). The agricultural sector has played a key role in Chile's economic success, both benefiting from stability and reforms and making an important contribution via rapid export growth. The main agricultural products with potential (fruits, wine, forestry, livestock, dairy) were identified early on, and policies and systematic programs were launched to raise productivity. The policies included long-term credit, technical assistance, public investment in basic processing infrastructure, and incentives to organize commodity cooperatives (OECD 2008; World Bank 2020). Agriculture responded with a high annual rate of production of 5 percent (FAO 2009) and has averaged about 4 percent (2010–18).

A policy focus on agricultural research and development, education, and extension has strengthened over the years, resulting in a diverse agricultural

(box continues next page)

Box 6.3, *continued*

innovation system (AIS) entailing significant public-private collaboration. Since the early 1980s, support to research and development has increased steadily, to about 10 percent of total budget transfers to agriculture, standing at 1.65 percent of agricultural GDP in 2013 (Stads et al. 2016). The initial focus was on adaptive research—Chile incorporated much foreign technology and knowledge, which played a significant role in, for example, fruit, wine, and salmon sector development (Chandra 2006). Today, the generation and adoption of agricultural technology involves a diverse set of institutions, foundations (for example, Fundacion Chile), universities, and companies operating with a high degree of autonomy.

Agri-food and forestry sector productivity has since leveled off, and concerns were raised over how to improve the profitability of the agri-food sector. Access to technology was expected to become increasingly difficult and expensive in the future (because of increased intellectual property rights). Climate change was expected to constrain agricultural development and international trade opportunities. Chile's authorities were keenly aware of the continued need for new technologies, new organizational and institutional arrangements, and enhanced farmer skills to successfully confront the future (World Bank 2011).

In 2010, the government, together with a wide group of stakeholders and the World Bank, engaged in a series of consultations and analysis (for example, scenario analysis, vision and strategy development, investment plans) to support the development of a long-term agricultural innovation strategy. According to its vision, in 2030 Chile will be a quality producer of a range of food and fiber products. Its international image will be marked by the diversity that its geography allows it to produce. The sector will emphasize environmental sustainability and wholesomeness, valued by both domestic and international consumers. Through the application of information and communication technology, investments in agricultural technology, and the training of its labor force, Chile will be able to develop profitable value chains, well integrated from production to final markets, and to remunerate its participants at comparable levels to the rest of the rural economy (World Bank 2011).

The jointly developed Vision 2030 guided subsequent coordination, governance, and investment efforts. The following main areas of development were pursued: strengthening Chile's AIS in comparison with other Organisation for Economic Co-operation and Development countries, strengthening the availability of new information and knowledge to producers, improving technological control over production systems, enhancing quality compliance and certification systems, and improving the human resource base, especially within the value chains (World Bank 2011).

Building leadership in AIS was considered essential to make the shift from being an adopter country to being an inventor. Although innovation was clearly prioritized at the national level, increased funding was available, and the AIS was diverse with modern characteristics, some key aspects of a strong AIS were missing, especially leadership, strategy, and facilitation. To enhance such leadership and confidence, a few actions were pursued:

- Support was to be provided to the Ministry of Agriculture in its capacity to manage the issues related to agricultural innovation by the establishment of a Directorate for Innovation whose main responsibility would be to ensure the participation of the sector in the national innovation system (NIS) and facilitate the implementation of its own agenda within the sector.
- Support the Directorate for Innovation in developing a strategy to articulate the positions of the agriculture sector within the NIS, thereby contributing to the strengthening of the NIS in general.
- Enable the Ministry of Agriculture (by making funds available to national producer and agri-industry associations for the first two years) to invite the private sector to strengthen its organization at the sector and key subsectoral levels (Chile Action Plan 2030).

INCREASING AGRICULTURAL R&D FINANCING TO SUPPORT DEVELOPING EAST ASIAN COUNTRIES' RESPONSE TO EMERGING NEEDS

Long-term growth in agricultural total factor productivity requires investments in agricultural research and extension. The impact of investments in agricultural R&D is often found to outpace the impact of other critical public goods, for example, roads, irrigation, electricity, or education (Mogues et al. 2012). In agriculture, R&D affects output with a long lag, but the impact lasts for a long time (Alston 2010). Research also demonstrates that countries that invested the most in R&D while simultaneously investing in extension have had the strongest productivity leap (World Bank 2005). Efforts to enhance the efficiency of the AIS should focus on reducing lags between R&D efforts and adoption of agricultural innovations. Agricultural R&D spillovers tend to be geographically bounded because innovations produced in one part of the world require adaptations to work well in local soil and climate conditions, underscoring the importance of domestic R&D efforts. Public spending in agricultural research (table 6.2) largely omits many crucial aspects of the innovation process: prioritization of investments, role of the private sector, public-private partnerships (PPPs), collective action and governance, and an enabling environment. However, overall public agriculture R&D is a reasonable proxy for a country's readiness for innovation that benefits society.

In China, current agricultural R&D spending and allocation may limit the country's ability to respond to the agri-food system's existing and emerging needs. Since 2000, most of the growth in agricultural spending in the region, which more than doubled between 2002 and 2010, has been driven by China, which has the world's largest and most relatively well-funded agricultural R&D system (box 6.4) (Chen, Flaherty, and Zhang 2012). The intensity of agriculture R&D spending (total agricultural research spending as a share of agricultural GDP) in China remains lower (at 0.62 percent) than in most Organisation for Economic Co-operation and Development (OECD) countries and in Brazil (average across countries, 2.51 percent) (OECD 2016) (table 6.2). However, in absolute terms, China's agricultural research spending far exceeds that of any other country except the United States (Chen, Flaherty, and Zhang 2012). Spending in agricultural R&D has not kept up with sector growth as well as R&D in other sectors.[9] China has increased spending particularly in biotechnology (box 5.5), including genetic modification and gene editing. In 2015, investment in the crop sector accounted for 56 percent of public agricultural R&D, and livestock accounted for about 10 percent (OECD 2016), suggesting that the past emphasis on grain self-sufficiency still drives the R&D agenda at the expense of food safety and zoonotic and environmental concerns.

Malaysia is the only country meeting the recommended 1.0 percent of agricultural GDP spent on R&D (table 6.2). However, its R&D spending has fallen significantly since the early 2000s. Malaysia's national agriculture research system is anchored by large national agricultural research institutes, complemented by several smaller government and higher education agencies (Flaherty and Dardak 2013). The palm oil, rubber, and cocoa commodity boards play a particularly important role in agricultural research. The predominant crop under research is oil palm (34 percent of all crop research in 2010, the country's primary export crop) (Flaherty, Stads, and Srinivasacharyulu 2013). Malaysia's R&D sector has an overall strong commodity focus, and relatively little R&D has gone to food crops, especially non-rice crops.

TABLE 6.2 **Parameters of agricultural research and development**

COUNTRY OR GROUP	SPENDING IN AGRICULTURAL R&D		SOURCE OF RESEARCH FUNDING AMONG PUBLIC R&D[b]			CO-INNOVATION AND MARKET-ORIENTED ACTIVITY[c]	PRESENCE AND PARTNERSHIP WITH IAR CENTERS	MAIN RESEARCH TOPICS AND BUDGET ALLOCATION
	PUBLIC (% OF AGRICULTURAL GDP)[a]	PRIVATE (% OF TOTAL R+D SPENDING)	GOVERNMENT	DONOR	SALES, SERVICES, LEVIES			
China	0.62	25	Most	—	Sales (41% in 2005)	CRGs, PPPs, firm acquisitions, and venture capital Special economic zones with tax incentives Patenting and TTOs	AVRDC, Bioversity, CIP, CIMMYT, ICRISAT, ILRI	Biotechnology, crops (56%), livestock (10%), limited cross-cutting
Indonesia	0.31	19	Main	—	Sales, contracts, commodity levies to be reinstated	CRGs, PPPs, contract research (universities) Patent licensing through TTO National funding schemes (centers and universities)	Bioversity, CIFOR, CIMMYT, CIP, ICRAF, IRRI	Staple crop and commodity focus, biotechnology, CSA, limited cross-cutting
Lao PDR	0.42	Minimal	< 50%	55%	Sales and services (1%)	—	Bioversity, CIAT, ICRISAT, ILRI, IRRI, IWMI	Crops (50%), livestock (about 17%), limited cross-cutting
Malaysia	1.0	High	Main	—	Commodity levies (27–78%)	CRGs, PPP in palm oil and rubber, tax incentives for private R&D	CIMMYT, WorldFish	Commodity focus (about 60%), livestock (about 12%), limited cross-cutting
Myanmar	0.04	Minimal	Most	High	—	Inclusive priority setting	CIMMYT, ICRAF, ICRISAT, IRRI, IWMI	—
Philippines	0.14	12–18	Main	5%	—	Tax incentives for private R&D Local partners with R&D	AVRDC, Bioversity, CIMMYT, CIP, ICRAF, ICRISAT, IRRI	Rice and commodity focus, biotechnology, limited cross-cutting

(continued)

TABLE 6.2, *continued*

COUNTRY OR GROUP	SPENDING IN AGRICULTURAL R&D		SOURCE OF RESEARCH FUNDING AMONG PUBLIC R&D[b]			CO-INNOVATION AND MARKET-ORIENTED ACTIVITY[c]	PRESENCE AND PARTNERSHIP WITH IAR CENTERS	MAIN RESEARCH TOPICS AND BUDGET ALLOCATION
	PUBLIC (% OF AGRICULTURAL GDP)[a]	PRIVATE (% OF TOTAL R+D SPENDING)	GOVERNMENT	DONOR	SALES, SERVICES, LEVIES			
Thailand	0.42	18–30	Most	—	Contract research	PPPs involving many actors Contract research Demand-driven R&D is rare Large firms with close contacts with farmers, research, and universities Tax and nontax incentives for private R&D Science parks with incubation services	AVRDC, CIMMYT, ICRAF, ICRISAT, IRRI	Biotechnology, livestock, rice, limited cross-cutting
Vietnam	0.19	10	Most (67%)	9%	Sales and services (10%) levies, producer organization (7%)	Tax incentives for private R&D Producer organizations pay for R&D New funding schemes	Bioversity, CIAT, CIMMYT, CIP, ICRAF, ICRISAT, ILRI, IRRI	Crops (35%), livestock (13), commodities (about 40%), CSA, limited cross-cutting
EAP Region	0.60	—		—			—	—
OECD and Brazil (average)	2.51	50		—			—	—

Sources: Original table for this publication, drawing on Aquino et al. 2014; Agricultural Science and Technology Indicators Database (International Food Policy Research Institute), 2020; Flaherty and Dardak 2013; Flaherty, Stads, and Srinivasacharyulu 2013; Fuglie et al. 2020; Mogues et al. 2012; OECD 2017b; Stads, Haryono, and Nurjayanti 2007; Stads and Kam 2007; Stads and Manivong 2006; Suphannachart 2017; World Bank 2017b.

Note: — = data not available or appropriate; AVRDC = World Vegetable Center; Bioversity = Bioversity International; CIAT = International Center for Tropical Agriculture; CIFOR = Center for International Forestry Research; CIMMYT = The International Maize and Wheat Improvement Center; CIP = International Potato Center; CRG = competitive research grant; CSA = climate-smart agriculture; EAP = East Asia and Pacific; IAR = International Agriculture Research; ICRAF = World Agroforestry; ICRISAT = International Crops Research Institute for Semi-arid Tropics; ILRI = International Livestock Research Institute; IRRI = International Rice Research Institute; IWMI = International Water Management Institute; OECD = Organisation for Economic Co-operation and Development; PPP = public-private partnership; R&D = research and development; TTO = Technology Transfer Office.

a. China (2010), Malaysia (2010), Thailand and Indonesia (2008), Vietnam (2015), EAP (2000–11).

b. Data not always available and may be outdated.

c. Commercialization through, for example, contract research, licensing agreements, TTOs, or incubators. Tax incentives for private sector research and services are also included. Co-innovation through, for example, CRGs, consortia, PPPs, participatory technology development, matching grant schemes, or science parks.

BOX 6.4

Drivers of agriculture sector growth in China

In China, complementary public investments into agricultural technology and the research-extension-training system have served as the primary engine of agricultural growth (along with the Household Responsibility System). Despite experiencing a twisting path of reform in agricultural research and extension, China has developed a strong research and development system. After the 1960s, China's research institutions grew rapidly, from almost none in the 1950s to a system that now produces a steady flow of new varieties and other technologies. During the 1980s and 1990s China's producers were replacing varieties from about 20–25 percent of their land during each cropping season (World Bank 2020).

The government provided all funding for research before the reforms were initiated in the mid-1980s. There has since been a gradual shift from formula-based financing to competitive grants. Agricultural research institutes also generate revenue from commercial activities, accounting for 41 percent of the budget (FAO 2005). Since the mid-1980s, public investment in agricultural biotechnology has increased significantly (FAO 2005), reaching more than $800 million in purchasing power parity terms (FAO 2006). Estimates of government support to agricultural research and development vary. In 2013, China's public funding of agricultural research was close to $10 billion, whereas in 2015 the estimates ranged from 26 billion to 55 billion renminbi (about $3.7–$7.8 billion, based on the June 2020 exchange rate). Over the past decade, both state-owned and other enterprises have engaged in agricultural science and technology activities in China (Huang and Rozelle 2018).

China's experience with special economic zones (SEZs) has developed over time, beginning in the early 1980s when market-oriented reforms were introduced in selected SEZ areas to attract foreign direct investment (FDI), followed by the establishment of open coastal cities, high-tech development zones, and later diverse regional zones. The SEZ incentives for stimulating growth and FDI entailed, for example, tax incentives for foreign investments, greater independence from the central government, and a focus on primarily export-oriented products and market-led activities (CDB 2015). Today, SEZs vary in scope and function. Some are designated geographical spaces where special policies and measures support specific economic functions. Others include free-trade areas, industry parks, technical innovation parks, and bonded zones that facilitate experimentation and innovation in a wide range of industries. By 2014, China was home to 31 bonded areas, 114 national high-tech development parks, 164 national agricultural technology parks, 85 national eco-industrial parks, 55 national ecological civilization demonstration areas, and 283 national modern agriculture demonstration areas (CDB 2015).

FDI has been one of the most significant features of China's economic reform and opening to the outside world. China became the world's largest recipient of FDI in 2004 (FAO 2006), with agriculture's share of FDI standing at 2.33 percent (1997–2014). The large volumes of FDI have led to more efficient resource allocation, rapid dissemination of new agricultural technologies, and the widespread application of agricultural mechanization (Chen 2018). China also has significantly increased its own net FDI to agriculture, from $190 million in 2006 to $3.29 billion in 2016, an average annual increase of 33 percent (Jiang, Chen, and Wang 2018).

Thailand's R&D spending is surprisingly low (0.42 percent of agricultural GDP) and heavily concentrated around the Greater Bangkok area (UNCTAD 2018). This concentration has resulted in limited co-innovation and diffusion of innovations to farmers and entrepreneurs. The bulk of the research falls under the Ministry of Agriculture and Cooperatives, which oversees four main research departments that focus on rice, other crops, livestock, and fisheries. The country's universities, Kasetsart University in particular, play a critical role in agricultural research (OECD 2017b). The lack of R&D coordination is a concern; there is no central policy or mechanism for coordinating R&D programs or their impact, nor is agriculture STI well integrated into the overall STI policy (OECD 2013, 2017b; UNCTAD 2018).

Indonesia has significantly increased its spending on R&D and has a well-staffed and relatively well-funded agricultural R&D system (Stads, Haryono, and Nurjayanti 2007; OECD 2017b). Most agricultural researchers are employed in a small number of government research agencies.[10] The higher education institutions (HEIs) (dominated by Bogor Agricultural University) are also involved in agricultural research. The government's interests in food self-sufficiency in staple crops and investments in palm oil plantations have largely driven R&D to staple and plantation crops at the expense of horticulture and livestock. In many areas of the country, however, farmers still grow rice varieties that were released 30 years ago (IFPRI 2019). As a long-term strategy, Indonesia aims to achieve a competitive position in the global biotechnology market (OECD 2017b). However, this vision and reality do not yet match (box 6.5).

The Philippines is currently underutilizing its vast but inadequately coordinated and funded R&D system. By number of staff, the Philippines has one of the largest agricultural R&D systems in Asia, but public expenditures and efficiency have been low. Agricultural R&D spending increased significantly between 2010 and 2013 but is still exceptionally low (0.14 percent of agricultural GDP) (table 6.2). Although domestic spending in R&D is low, it is partly compensated for by international collaboration (see the section titled "International collaboration for a regionwide response to agri-food system challenges" in this chapter). The Philippines has a large, complex (hundreds of HEIs, research centers, and centers of excellence), and fragmented agricultural R&D system that has evolved over the years and through changes of government. Coordination has been improved, and today the Philippine Council for Agriculture, Aquatic, and Natural Resources R&D acts as the central coordinating body providing support to 132 implementing R&D agencies (OECD 2017b). With regard to focus, although more than half of agricultural research agencies conduct research on crops, R&D on rice is privileged (OECD 2017a). The pressing needs call for greater R&D in climate-smart agriculture, e-extension, precision agriculture, and the food chain.

BOX 6.5

Biotechnology research and development in Indonesia

Indonesia established Inter-University Centers in Bogor Agricultural University, Bandung Institute of Technology, and Gadjah Mada University in Yogyakarta focusing on agricultural, industrial, and medical biotechnology, respectively. These centers are attached to the universities and play an important role in the development of biotechnology in Indonesia in their respective areas (OECD 2017b).

The government and local universities continue to conduct research into several gene-edited crops, albeit at a moderate pace. Research includes virus resistance for tomato, rice, potato, and sugarcane. In addition, Indonesian researchers are genotyping Indonesian livestock, including poultry, sheep, Balinese cattle, and fish. Additional research being carried out includes identification of rapid growth and disease resistance in catfish and common carp. Practical applications in Indonesia for gene-edited animals and animal products remain far into the future.

Sources: OECD 2017b; USDA 2018.

Vietnam significantly underinvests in agricultural R&D, putting the agri-food sector's long-term development at risk (table 6.2). The misalignment between Vietnam's strategy and high-tech aspirations (box 6.6) with its current investment in agricultural R&D (0.19 percent of agricultural GDP) and

BOX 6.6

Vietnam's vision for greener high-tech growth

Agricultural research and development (R&D) is still a peripheral activity in Vietnam. Over the years, Vietnam has made modest investments into agricultural technology, inputs (for example, subsidized fertilizer, improved varieties), and services, all of which played a significant role in the sector's take-off (World Bank 2016). Various research centers were established in the 1980s. By 2003, 28 different research agencies had been placed under the Ministry of Agriculture and Rural Development. Public agricultural R&D expenditures nearly tripled during 1996–2002. Public investment into research totaled $71.3 million in 2002, that is, 0.88 percent of agriculture, fisheries, and forestry value added (Stads and Hai 2006; World Bank Databank). Vietnam's investment into agricultural R&D has since declined significantly, from 0.88 percent in 2002 to 0.19 percent in 2015.

Private agricultural R&D is limited. Vietnamese enterprises have been reluctant to develop in-house R&D. Most enterprises rely on imported technology in the form of machinery and equipment, use technology transfer from abroad, or joint venture with foreign partners (Hoang 2017). Since 2014, private R&D has gradually increased. Viettel Group has taken the lead in science and technology (S&T) investing (10 percent of its expenditure). It spent $200 million on S&T and has established three research institutes. Thai Binh Seed Company set up an R&D center for new products with a scale of tens of hectares, investing hundreds of billion Vietnamese dong for research projects. Thanks to strong investment in S&T, the company has successfully bred hundreds of new hybrids, testing thousands of new plant varieties. The strong investment in S&T has made Thai Binh Seed Company one of the leading companies in the seed industry in Vietnam. The evidence indicates that, once an enterprise is properly aware of the role of S&T and transforms it into action, S&T can be vigorously developed, contributing to the development of enterprises (World Bank, n.d.).

Vietnam is in pursuit of innovation and emerging technologies for agriculture and economywide transformation by 2050. The government has identified four priority industries: emerging technologies (artificial intelligence, robotics, and smart systems), education and training, high-tech agriculture and food processing, and green energy, environment, health, and tourism. Of these, emerging technologies is the most important given that it supports the other sectors and sets the tone for Vietnam's ecosystem development (Austrade 2019). Among the emerging technologies, interest in blockchain is growing rapidly. Indeed, Vietnam has the potential to become Southeast Asia's blockchain innovation hub (Hynes 2018).

To improve yields, overall productivity, and the sustainability of agriculture, Vietnam is in the process of developing 10 high-tech agricultural parks and zones (2020–30) with investment in R&D, extension, training, demonstration, and processing of several high-value commodities (Official Gazette 2015).

Vietnam's largest cities are also interested in emerging technologies. The local government in Da Nang is focused on boosting growth in tourism, high-tech industries (information and communication technology and software development), and the logistics sector. In Ho Chi Minh City, priority sectors include high-tech (that is, precision engineering, automation), micro-electronics, information technology, telecommunications, biotechnology applied to pharmaceuticals and the environment, new materials, new energy, nanotechnology, agritech, edtech, the Internet of Things, and food processing.

Vietnam is also striving to improve the enabling environment and has invested in key organizations, provided support to start-ups, and improved training efforts. Despite endeavors to improve the ecosystem for emerging technologies, Vietnam still faces several main challenges, such as limited access to finance, talent and entrepreneurial skills, R&D investment, and management of intellectual property rights.

overall AIS is wide. In 2010, the research allocation was rather commodity oriented, accounting for about 50 percent of all agricultural research. Only 35 percent of scientists focused on crop research and livestock accounted for about 12 percent (Flaherty, Stads, and Srinivasacharyulu 2013). The scattered and small-scale nature of funding has led to poor quality research compounded by the relative absence of research evaluation (OECD and World Bank 2014). The R&D human resource base is relatively thin in Vietnam (Flaherty, Stads, and Srinivasacharyulu 2013; OECD 2017b). The structure of public agricultural R&D is complex, with multiple government entities and HEIs engaged in agricultural research. The institutional setup has, however, undergone significant changes since 2010. The Vietnam Academy of Agricultural Sciences currently oversees the bulk of the country's agricultural research (OECD 2017b).

A weak human resource base presents a major bottleneck for innovation in Cambodia and Lao PDR. The national agricultural research systems in Cambodia and Lao PDR are anchored by large national agricultural research institutes, complemented by several smaller government agencies and HEIs. Both Cambodia and Lao PDR still have very low numbers of PhD-qualified staff, posing a significant impediment to advancing the quality of research. Nonetheless, these countries have wisely invested heavily in staff recruitment and training in recent years (OECD 2013, 2017b; Stads and Manivong 2006). They also suffer from outdated equipment and facilities that impede productive research and compromise the number and quality of research outputs.

Lao PDR has been able to increase its agricultural R&D spending (0.42 percent of agricultural GDP); however, spending is low in Cambodia (0.20 percent of agricultural GDP) and dismal in Myanmar (0.04 percent of agricultural GDP) (table 6.2). The three countries may continue on their current path of adopting existing "traditional" and new innovations (for example, climate-smart agriculture, genetically modified [GM] crops, e-extension), as well as continue to rely heavily on international agricultural research centers for new productivity- and sustainability-enhancing innovations and on donors for funding.[11] However, should they continue to underinvest in their domestic R&D capacity, they may jeopardize their agri-food sector development.

PROVIDING THE INCENTIVES AND BREAKING THE BARRIERS TO INCREASE PRIVATE SECTOR R&D

At the global level, the private sector has assumed a large role in developing improved technology for the agriculture and food sectors. Major drivers have been new commercial opportunities afforded by scientific advances and liberalization of agricultural input markets. In 2008–09, global public agricultural R&D expenditures stood at $32 billion to $34 billion; private agricultural R&D spending worldwide more than tripled between 1990 and 2014 to $15.6 billion per year (Bientema et al. 2012; Pardey et al. 2015). Although most private R&D is conducted by a relatively small number of companies and in high-income countries (88 percent), a significant portion is likely targeted toward developing-country farmers (estimated at 28 percent for the top tier companies) (Fuglie 2016).

The rising importance of private R&D does not imply a diminished role for the public sector. Rather, the role of public research changes along with agri-food system development. Most empirical evidence points to complementarities between public and private agricultural R&D. It is essential that the public sector continues to invest in scientific capacity, training, and skills, and overall, in innovations with limited private sector interest (for example, smallholders in remote areas, climate-smart agriculture, improved natural resource management, maintenance of biodiversity, zoonoses) that typically fall outside the scope of purely private innovations. The public sector, in concert with international organizations, also needs to ensure predictable and consistent financial support to basic research. It must also be conscious of pursuing synergies and avoiding duplication of efforts to improve the overall efficiency of the use of financial resources (Group of 20 2012). However, the public sector could crowd out or regulate out private investment in agricultural R&D if it competes with private investors in the development of new technologies.[12] Such crowding out can be avoided if the public sector focuses on the areas in which a large gap between social and private returns on investment exists (OECD 2018a).[13] Box 6.7 discusses the shifting role of public and private agricultural research (both basic and applied) in transforming and urbanizing contexts.

BOX 6.7

The shifting role of public and private agricultural research and development in transforming and urbanizing countries

Public agricultural research has continued to wrestle with the issue of how to be more responsive to demand and how to balance farmers' needs with improved consumer acceptance. Private research capacity, however, develops principally where agricultural markets function well or where specific incentives exist to address market failures.

Frontier science. As agricultural economies modernize and the private sector becomes more active in funding its own research, innovation turns more to the application of frontier science, and public research tends to support private companies by developing new products (for example, hybrid rice) or supporting private sector research. In agriculture, molecular biology and genomics represent this kind of frontier science, which is often supported through competitive grant schemes and in which universities often have a comparative advantage. For example, India's National Agricultural Innovation Project (NAIP) operates a competitive grant scheme that funds innovation clusters around more basic research with potential applications of interest to the private sector. In Thailand, similar efforts are led by the Ministry of Science and Technology through its National Innovation Agency and BIOTEC program, which also focus on funding clusters of research and related applications. In addition, BIOTEC has set up two independent research programs, the Rice Gene Discovery Unit and the Cassava and Starch Technology Research Unit. In these cases, public sector research is increasingly divorced from farmers as the primary clientele, relying instead on input markets as the mechanism for articulating farmer demand.

Institutional innovations. Under these market-driven conditions, investments in public agricultural research tend to focus more on institutional innovations that reinforce the ties between research and the private sector. Intellectual property rights are emphasized to ensure open access to publicly generated innovations and to protect innovations developed in the private sector. Intellectual property rights are often the basis for contractual arrangements in public-private partnerships. This connectivity can be reinforced by competitive grants that insist on public-private partnerships, brokers that can mediate between public research and subsector needs, science parks adjacent to research institutes that focus on areas of joint research and development, and venture

(box continues next page)

Box 6.7, *continued*

capital funds that invest in developing products and markets based on research innovations.

Applied research with public interest. Other areas of applied agricultural research are less well served by the private sector and constitute more classical public goods, such as pulse and grain legume breeding, crop and livestock disease surveillance, development of forages for ruminants, and especially crop management and natural resource management research. Agricultural research institutes within dynamic agricultural sectors have to find a balance between the more basic research that complements the private sector's interests and the more applied research that farmers need. This balance will become even more important with the increasing focus on using water and nutrients efficiently and reducing environmental externalities in production systems.

Lagging regions. Even transforming agricultural economies need public sector research in lagging rural areas with high poverty rates, usually associated with underdeveloped markets. Demand articulation and connectivity usually focus on brokering organizations, particularly nongovernmental organizations and extension services. India's NAIP is an example of a funding program that stratifies its platforms or clusters based on relative market development and associated rural poverty. In more commercialized areas, NAIP's platforms involve public-private partnerships, and in lagging areas they involve traditional research and brokering organizations. For smaller research institutes, this kind of stratification creates a dilemma. Should they focus on the more commercial areas and associated partnerships or focus on the lagging areas? The potential for innovation will be higher in the commercial areas, but the public interest may reside with the lagging areas. The tendency within an agricultural innovation system will be toward the former, whereas the public role will in most instances lie in the latter.

Source: World Bank 2012.

The private sector plays a growing role in the region in raising the agri-food system's technology level and increasing farmer access to high-tech applications and services.[14] The private sector mainly invests in agricultural R&D activities that have a high market return on investment. They tend to concentrate on applied and development research and on areas that protect their returns through intellectual property rights (IPR). For example, private investors favor sectors such as food processing, agricultural chemical inputs, farming machinery, hybrid seeds, and GM crops breeding. The private sector has also often been in the driver's seat regarding new transformative innovations (for example, precision agriculture, e-services, blockchain applications).

Private sector R&D spending is on the rise in the region but is still below the average (50 percent) of the OECD countries (table 6.2). Private R&D accounts for about 25 percent of all agricultural R&D spending in China, Malaysia, and Thailand. However, private R&D accounts for more than 90 percent of all sector R&D expenditure in China (OECD 2016), suggesting that the agriculture sector imposes significant barriers or disincentives to private R&D.[15] Malaysia has significant private sector R&D, partly financed by levies, but most is directed toward the commodity sector (Flaherty and Dardak 2013). Thailand has among the highest levels of entrepreneurship in the world (OECD 2013), also reflected in the amount of private agricultural R&D, particularly by large processing firms.[16] In China, Malaysia, and Thailand, private investment in agricultural R&D is undertaken by enterprises, financial institutions, and venture capitalists (mostly nascent; see the section titled "Building private sector capacity for market-based innovation is necessary for agri-food system transformation" in this chapter). The public and private sectors also jointly invest in agriculture R&D (box 6.8).

Private agricultural R&D is also increasing in entrepreneurially active Indonesia and the Philippines. Indonesia is a frontrunner among its peers regarding private R&D, co-innovation, experience with associated instruments (for example, competitive research grants, PPPs, contract research), and commercialization of innovations (for example, patent licensing) (Hall et al. 2016; OECD 2013, 2017b). Indonesian agencies are actively searching for PPP models (box F.6 in annex F), yet the activity can be considered nascent. The plantation companies account for most of the private R&D (close to 60 percent) (Stads, Haryono, and Nurjayanti 2007). The private sector (mostly firms and private HEIs) also plays a relatively important role in conducting agricultural R&D in the Philippines. Private sector R&D primarily addresses plantation crops such as bananas and pineapples, agrochemicals, pest management, and plant and livestock breeding (OECD 2017a, 2017b).

The transformation of Vietnam's agri-food system is hampered by limited engagement of the private sector. To date, private sector activity and private R&D—by state-owned enterprises, small and medium enterprises, and foreign firms—have played a limited role (box 6.6), focused mostly on plant varieties, agrochemicals, and production technologies. Despite the importance of

BOX 6.8

Collaboration models between public agricultural research and development and the private sector in China

In China, public research and development (R&D) institutions have been collaborating with the private sector since the early 2000s to complement research funding and to apply R&D outcomes in practice. In general, five categories of collaboration exist. Restrictions apply to collaboration with foreign firms, but acquisition of firms has served as a way to address this challenge.

1. Research staff in public R&D institutions engage in R&D activity in private enterprises on a part-time basis. This is one of the simplest and most feasible modes of cooperation.
2. Joint development of new technology has become a common model of cooperation: public R&D institutions, higher education institutions, and enterprises work together to develop new materials, products, technology, and equipment.
3. Indirect cooperation between public R&D institutions and the private sector through intermediaries such as brokers, consulting firms, industry associations, federations, and government agencies and departments has become increasingly common. These intermediary institutions play a major role in bridging, monitoring, and coordinating the cooperation.
4. Public R&D institutions and the private sector sometimes establish joint research institutions. This model of cooperation can clarify the direction of R&D and reduce the waste of resources, thus shortening the research cycle. It also can spread the responsibility between researchers and the private sector, thus shortening the industrialization cycle of R&D outcomes.
5. Public R&D institutions and the private sector could set up enterprises. In the current legal framework, three types of companies are the most practical: limited liability companies, joint stock limited companies, and cooperative organizations.

Sources: Fuglie 2016; OECD 2018a.

agricultural exports, value added in the export commodity sector is limited (World Bank 2016). However, Vietnam emphasizes innovation, for example, in emerging technologies and high-tech agriculture and food processing, with incentives for greater private R&D and overall investment. Private sector investment in all sectors is impeded by limited access to finance, talent and entrepreneurial skills, and management of IPR (Austrade 2019).

The countries exhibit much heterogeneity, owing to their respective states of development and the priority given to R&D and innovation. Most MICs, including those in developing East Asia aspiring to build stronger AIS, have imbalances in their existing AIS. The countries have largely public R&D institutions, but private sector–led R&D activity is low. Furthermore, domestic public R&D institutes and HEIs often have little interaction with businesses, which in turn exert little demand for their R&D-related services. Thus, demand-led R&D is still limited in the region, and stakeholders have relatively little experience in competitive research grants, matching grants, and research-innovation consortia[17]—all common instruments that foster co-innovation (table 6.2). Considerable room still exists for greater involvement of the private sector and other actors, such as cooperatives, farmer groups, and extension, in innovation processes. The size and organizational arrangement of R&D systems vary greatly across countries. In many countries, large numbers of scattered research organizations and HEIs or lack of overall coordination of R&D have resulted in inefficiencies in resource use. Increasing coordination of innovation and R&D institutions at the national level and the strengthening of evaluation mechanisms have, however, improved the efficiency of public funding to agricultural R&D.

Developing East Asian countries' agricultural R&D focus is not suitable for addressing existing and emerging needs. There are significant cross-country differences in R&D allocation (table 6.2). Data on public R&D spending are limited or outdated in many countries; however, reports suggest that public R&D spending targets staple crops (especially rice) and commodities (Flaherty, Stads, and Srinivasacharyulu 2013; IFPRI 2019; OECD 2016, 2017a). Relatively little research is carried out on livestock (about 3 to 17 percent), fruit and vegetables, and post–farm gate activities (for example, processing, food safety) and multidisciplinary issues (for example, safety, nutrition, zoonoses, natural resource management, socioeconomic research). Consumer preferences, the poor status of the agro-ecological system, persistent food safety and nutrition challenges, and accelerating zoonoses indicate that public agricultural R&D allocation urgently requires adjustment to better accommodate emerging needs.

INTERNATIONAL COLLABORATION FOR A REGIONWIDE RESPONSE TO AGRI-FOOD SYSTEM CHALLENGES

A country's ability to generate new innovations depends crucially on its capacity to absorb and build upon existing ideas, most of which are often foreign. Thus, international knowledge links are central to the development of AIS and are often viewed as a catalyst for sector development. Such foreign technology and knowledge links may include FDI, licensing, international co-invention, labor migration, and international trade. Nonmarket interactions, such as scientific collaboration and aid from governments in the form of development assistance, also play an important role in technology transfer and in capacity building (OECD 2013).

The benefits of international cooperation for AIS stem from the specialization it allows and from international spillovers. It is particularly important where global challenges (for example, responding to climate change) or transboundary issues (such as water use, pest and disease control, zoonoses) are confronted, and when initial investments are exceptionally high (Group of 20 2012). Cross-country cooperation also allows countries to better leverage their domestic research resources, which may be particularly important for small countries with limited domestic R&D capacity. Thus, domestic investment in agricultural R&D does not convey the whole picture, given that many countries either have a prominent presence of International Agricultural Research Centers or have strong strategic partnerships with them (table 6.2). The Philippines, for example, hosts the International Rice Research Institute,[18] and Indonesia is home to the headquarters of the Center for International Forestry Research and World Agroforestry and leverages significant international R&D. Strengthening scientific and knowledge exchanges with centers of excellence (CoE) in neighboring countries is also an important element of an innovation strategy for a small country with limited resources. However, relying on international R&D capacity may not be sustainable as a long-term strategy. All countries also collaborate with bilateral and multilateral development agencies and international research networks. The collaboration networks have centered largely on topics that receive less domestic attention (for example, livestock, farming systems, the environment) or pose significant cross-country challenges. For more details on international collaboration among the countries, see appendix F.

FDI is one of the most important channels through which technology is transferred across countries. FDI reflects a country's level of integration into international knowledge networks. Some countries' AIS have attracted considerable FDI. For instance, in China, FDI and the associated special economic zones have contributed significantly to innovation through transfer of new agricultural technologies, knowledge, and finance (CDB 2015; Chen 2018) (box 6.4). China has also accessed foreign agricultural technology through direct acquisition of foreign firms. This strategy may speed up transfer of technology assets to China and preserve greater sovereign control but is financially costly to the Chinese economy (box F.1 in appendix F) (Fuglie 2016). Foreign research also plays an important role in transferring technology or knowledge to research agencies in Thailand (Suphannachart 2017). In Indonesia, FDI has been modest and volatile, and mostly concentrated in food and plantation crops. Although Cambodia and Myanmar enjoy relatively high FDI, limited R&D is involved (ADB 2019). However, foreign investment in agriculture and agricultural R&D is impeded in the region by inadequate investment conditions. Such conditions include, for example, infrastructure, finance, IPR, seed and biosafety regulations, land rights, and curbs on the use of FDI in agriculture (ADB 2019; Austrade 2019; OECD 2016, 2017b, 2018a).

RETURNS TO INNOVATION INCREASED BY REFORM OF AGRICULTURAL EXTENSION SERVICES AND INTEGRATION OF E-EXTENSION

Investing in extension services has paid off. Traditionally, agricultural extension services have facilitated knowledge flows from research institutes to farmers and entrepreneurs. Often, however, agricultural extension has been forgotten and underfunded, with poor delivery limiting the potential of agricultural innovations. Although evidence of the impact of some major extension models has been mixed, the median rates of return range between 58 percent and 63 percent for investments in extension services (Alston et al. 2000; Dercon et al. 2008). Extension services have also yielded positive social returns, particularly for women, people with low literacy levels, and farmers with medium landholdings (Davis et al. 2010).

Over the years, agricultural extension services have shifted away from public delivery of technology, inputs, and knowledge. In the past, the public sector dominated extension in most countries. Today, farmers are drawing information from an increasing range of sources and means, especially through innovations in information and communication technology (ICT). This pluralism in services is reflected in the range of approaches used, the array of content, and the interaction with public and private entities. New actors, including nongovernmental organizations, farmer organizations, the private sector, knowledge brokers, and community-based organizations, offer and fund services (World Bank 2012). The roles of these actors vary and are affected by the agricultural commodity, the level of market integration and value added, farmer education, farm size, collective action, and the country's views on the role of the private sector. As innovation processes have become more complex in increasingly market-oriented and challenging contexts, the need for neutral brokers has become more evident (box F.4 in appendix F).

ICT-based e-extension services have fundamentally changed the functioning of farm advisory services. They have improved timeliness and created new partnerships and forms of investment (chapter 5). Both publicly and privately provided extension services are more and more often delivered over mobile phones, which are increasingly multifunctional wireless devices, delivering timely and customized information at scale, in some areas resulting in yield increases of up to 4 percent (Fabregas, Kremer, and Schilbach 2019). The public sector maintains great interest in ICT as a means of providing better public services that affect agriculture (for example, land registration, forest management, and extension), as well as for connecting with citizens (World Bank 2017a). Private companies that have invested in technology and applications are often interested in working with the public sector to provide their products and services to smallholders. Mobile phone applications, software design, local language customization, and remote transaction services represent only a fraction of the opportunities for continued innovation.[19] Commercial enterprises such as processors, input suppliers, and exporters are also motivated to invest in ICT because these investments often lead to increased efficiency and revenue as well as extensions to client bases such as isolated farmers (World Bank 2017a).

Increasingly pluralistic (multiple-provider) extension services need new capacities to serve a varied clientele. Today, the clients—farmers and firms—need a variety of services to function in increasingly complex contexts. Paradigm shifts from the perception that research knowledge can drive innovation must also take place in the programs and among field staff, extension administrators, and policy makers (World Bank 2012). Service providers must be equipped with knowledge and skills in, for example, value chain approaches, market-oriented extension, climate change, risk management, group and organizational development, agribusiness, and mechanisms to share information, including use and provision of e-extension.

The roles of the public and private sectors in extension are also changing in most agri-food system contexts. The private sector (for example, services by agro-dealers and managers of out-grower schemes, private and third-sector providers) and knowledge brokers play an important role in providing services in contexts of transition and urbanization; however, government has a continuing role to play (Rivera and Alex 2004; Swanson and Rajalahti 2010). Many extension tasks have a public goods nature. Such tasks include regulation, quality control in the produce supply chain, coordination of service provision, and natural resource management, as well as the provision of services to marginal groups, which are unlikely to access or be able to afford private advisory services. The public sector's role is to fund the provision of extension services (directly or through outsourcing) where demand for services is not being met, to support extension services in addressing issues of long-term social and environmental sustainability, and to manage extension services (including quality control and knowledge management). The public sector can also provide incentives for non-public actors to play a greater role in providing services (box 4.5). In pluralistic extension systems, space can be created by the public sector to shift some public investment toward the management of extension systems and the strengthening of private actors' capacity (Christoplos 2010; Spielman et al. 2011).

The developing East Asian countries have largely public sector–driven extension systems with varying but increasing levels of pluralism, decentralization, and integration of e-extension (table 6.3). Pluralistic extension and rural advisory services prevail in Cambodia, Indonesia, Lao PDR, Thailand, and Vietnam (GFRAS 2016). The central challenge of extension systems is limited coverage and quality (technical, enterprise skills), particularly in Cambodia, Lao PDR, and Myanmar, which also have the poorest penetration of e-extension (with potential to improve quality and coverage). Extension systems are significantly underresourced, and some rely on donor funding in large part (for example, Myanmar). Limited decentralization still hampers smallholders' access to knowledge and innovations. For instance, Thailand's public, largely top-down, extension model does not adequately integrate client demand (GFRAS 2016). Some extension systems are hindered by content. For instance, the Philippines' extension system focuses mostly on a commodity approach (OECD 2017b), constraining climate-smart agriculture programs. Although extension is provided by many private actors, such as agro-dealers, processors (who provide a comprehensive package of inputs, advice, and credit to their contract farmers), and business-oriented cooperatives, technical skills also often limit private sector extension activities. As an example, China's nationwide extension system is featured in box 6.9.

Several developing East Asian countries have ambitious e-extension plans,[20] but integration into national extension systems varies (table 6.3). For instance, the Philippines' e-extension program aims to provide an alternative to the traditional extension system of the agricultural, fisheries, and natural resources sectors by integrating ICT-based extension delivery. The e-farming component delivers farm and business advisory services, primarily technical assistance, to farmers and agricultural extension agents via text and voice (using a toll-free number), as well as through chat, e-mail, and online forums[21]. Chapter 5 and appendix E provide details on e-extension in Vietnam and China.

Most developing East Asian countries have rather weak links between farmers, extension, and research (and firms), suggesting that farmer and firm demand is poorly informing R&D programs. In China Indonesia, Malaysia, and Thailand (at the local level), however, cooperatives and industry associations play a brokering role—to a varying extent—in facilitating collaboration, sharing knowledge, and promoting innovation. These organizations serve as platforms

TABLE 6.3 Agricultural extension coverage, level of pluralism, decentralization, information and communication technology integration, links between main actors, and key constraints

COUNTRY	AGRICULTURAL EXTENSION COVERAGE		LEVEL OF PLURALISM[c]	LEVEL OF DECENTRALIZATION	ROLE OF ICT IN EXTENSION	RESEARCH-EXTENSION-FARMER ORGANIZATION LINKS	KEY CONSTRAINTS
	NUMBER OF AGENTS (2009–12)[a]	RURAL POPULATION PER EXTENSION AGENT[b]					
Cambodia	1,244	10,000	High	Medium	Low-medium	Limited	Hierarchical, limited decentralization, coverage
China	617,706	920	Medium and increasing	High	High	Good (demonstrations, cooperative role, technology parks, private sector activity)	Despite improved coverage, not all households reached; quality
Indonesia	53,944	2,217	Medium	High	Medium-high	Through CRGs, industry associations, university R&D programs, institutes with community service mandate; however, not strong	Despite decentralization, limited attention to extension at lower government levels; quality
Lao PDR	752	6,103	Medium	High	Low	Limited	Coverage, quality
Malaysia	1,355	5,576	Mostly public, which also offers commercial services	Medium	Medium-high	Some, for example, through commodity boards	Limited pluralism
Myanmar	10,947	3,406	Mostly public	Medium	Low	Reasonable	Coverage, quality
Philippines	32,328	1,752	Mostly public	High	Medium-high	Some, for example, through Agricultural Training Institute coordination	Budget, fragmented structure, diminishing coverage, commodity-focused content
Thailand	18,881	1,840	High	High	High (e-learning, ICT channels, Smart Farm project)	Some, for example, through universities; inclusiveness for problem solving is rare Cooperatives have a liaison representative Informal links at local level are strong but at the national level weak	Limited research-extension links Top-down approach with limited quality and farmer access a barrier for diffusion—farmers using other channels
Vietnam	34,747	1,762	Medium but increasing	High	Low-medium	Some through open competitive bidding for delivery	Coverage, budget, limited pluralism (despite bidding)

Sources: Original table for this publication, drawing on Global Forum for Rural Advisory Services (GFRAS) World Wide Extension Study data set (https://www.g-fras.org/en/world-wide-extension-study.html) OECD 2018a; World Bank Databank, 2020 (https://databank.worldbank.org/home.aspx).

Note: CRG = competitive research grant; ICT = information and communication technology; R&D = research and development.

a. In 2015, there were 16,000 technical service centers, housing 729,000 extension officers who provided technical services to 12.5 million farming households, equivalent to a total of 60 million farmers in China (OECD 2018a).

b. Rural population from World Bank Databank (2020).

c. Role of public sector, private sector, nongovernmental organizations, farmer associations, and universities in providing extension (GFRAS, 2016).

BOX 6.9

Public agricultural extension system reforms in China

Agricultural extension was established in China in the 1970s. However, the commercialization of extension activities in the 1980s reduced its capacity to provide diverse services and advice. More recent reforms to improve the quality of service to farmers by separating commercial activity from extension services and introducing a more inclusive approach at the local level are more favorable. At the same time, private organizations are increasingly playing a major role in facilitating knowledge flows in China. For example, farmers' cooperatives often function as intermediate agents (such as with industry associations) to facilitate the adoption of innovations and reduce transaction costs, allowing smallholder farms to overcome systematic constraints to adopting innovations.

China also began an initiative to build agricultural technology parks to demonstrate new technologies and facilitate collaboration between agriculture and other industries. National Agricultural Science and Technology Parks are intended to create innovation hubs and an entrepreneurial chain to strengthen the transformation and incubation function of agricultural science and technology achievements. Each park is composed of a core area and a demonstration area. By the end of 2015, 246 such parks had been established using both public and private funds. This system also combines nonprofit and for-profit services, as well as special and comprehensive services (China, Ministry of Science and Technology 2016; OECD 2018a).

The public extension system should further evolve by strengthening decentralization, enhancing public good–type services (for example, environment, noncommercial crops, remote areas), and allowing greater pluralism in service provision. Greater use of e-extension, which is already under way, would also improve the performance of advisory services.

Sources: Global Forum for Rural Advisory Services World Wide Extension Study data set (https://www.g-fras.org/en/world-wide-extension-study.html); OECD 2016, 2018a.

for providing farmer feedback to research programs. These types of platforms and networks are crucial for fostering co-innovation, particularly for transformative innovations that often require human resources from many disciplines. Vietnam has also witnessed an increase in the number of knowledge brokers (box F.5 in appendix F).

BUILDING PRIVATE SECTOR CAPACITY FOR MARKET-BASED INNOVATION NEEDED FOR AGRI-FOOD SYSTEM TRANSFORMATION

The effectiveness of innovation systems depends on the depth and diversity of innovation capabilities that are accumulated by, and deployed in, innovation actors, including firms. Firms may generate new agricultural technology, adopt external technology, make minor adjustments to it, or use it in a new way (OECD 2013). Firms may also innovate via new business models and organizational arrangements (for example, value chain management, contract farming).

In general, firms would benefit from measures that improve their innovation capacity and access to finance. Firms can benefit from initiatives that facilitate networking and partnerships, which improve capacity. They address market failures at different stages of the networking process through firm-specific or less targeted measures.[22] Policies to reduce financing gaps fall into three broad categories: subsidized loans and loan guarantees, provision of seed financing and support for the development of venture capital, and tax incentives or grants to correct market failures that lead to underinvestment in R&D (OECD 2013). Often, measures to support firms to innovate combine both support to capacity and financing, as explained in the subsequent paragraphs.

Policies and instruments to foster greater private investment and inclusion of diverse stakeholders have been applied to varying extents in the region. As discussed in the section "Providing the incentives and breaking the barriers to increase private sector R&D," significant regulatory and institutional barriers impede firm innovation in developing East Asia. Several measures have been pursued to enable greater private investment and innovation, including R&D-based tax incentives, competitive research grants and support to PPPs, incubators (including IPR protection), and, to some extent, loans and risk capital. Networking support is not prominent; however, many instruments aim to engage diverse actors and thereby provide incentives for networking and partnerships.

R&D tax credits hold promise for overcoming market failures. More than two-thirds of OECD members provide tax incentives for R&D. Available evidence on the effectiveness of R&D tax credits is mixed, but they can be an effective mechanism for overcoming market failures resulting in underinvestment in private R&D (Hall and van Reenen 2000). Malaysia, the Philippines, Thailand, and Vietnam offer tax incentives for firms that undertake R&D and provide R&D services, which may also apply to the agri-food sector (box 6.10). Although Indonesia does not have an R&D-based tax incentive scheme, a range of tax incentives are available to Indonesian companies seeking new investments in Indonesia (Ernst & Young 2013; OECD 2017b).

Firms (and other innovators) can also be supported in the early stages to overcome the challenges of creating and commercializing innovations. See box 6.11

BOX 6.10

Research and development-based tax incentives for innovation

In *Malaysia*, companies that provide research and development (R&D) services are eligible for Pioneer Status (income tax exemption) or an investment tax allowance for qualifying R&D capital expenditures. A double tax deduction is available for R&D revenue expenditure incurred by companies carrying out in-house R&D or expenditures for the services of approved R&D service providers. There are also a variety of financial assistance schemes.

In the *Philippines*, R&D expenditures may be treated as a current expense, deductible at 100 percent, or a deferred expense ratably distributed over a period of not less than 60 months, as chosen by the taxpayer. Moreover, the 2012 Investments Priorities Plan identified R&D activities as investment priorities that promote the economic development of the Philippines. Enterprises engaged in R&D activities (as an R&D service provider) that qualify for registration with the Board of Investments may be entitled to a four-year income tax holiday and other incentives.

Thailand provides a 200 percent deduction for the cost of engaging approved Thai R&D service providers with no requirement for foreign-majority-owned companies to own the results of the R&D activities. Companies providing eligible R&D services may also be entitled to other incentives.

In *Vietnam*, newly established companies in high technology, science, and research domains are entitled to a reduced tax rate for 15 years, which can be extended to 30 years, subject to approval. There is also a one-year exemption for income derived from performing R&D, the sale of products during test production, and products made from new technology applied for the first time in Vietnam.

Source: Ernst & Young 2013; OECD 2017b.

for a discussion of incubators and other efforts. Technology transfer offices that offer, for example, IPR support for technology commercialization (for instance, Tsinghua University in China and Inova in Brazil [World Bank 2012]), can also be linked to incubator services.

BOX 6.11

Public-private efforts to stimulate greater private sector investment in innovation

Incubators. Incubators stimulate technology commercialization and business development. Some incubators are dedicated to accelerating technology commercialization or technology transfer. The former typically have strong ties with agricultural research institutions; oftentimes they are arms or spinoffs of such institutions. Examples include the International Crops Research Institute for Semi-arid Tropics–affiliated Agribusiness Incubator in India; IPB University in Indonesia; and Brazil's CENTEV/UFV Technology Incubator, affiliated with the Federal University of Vicosa. Villgro in India accelerates the uptake of indigenous technology with activities involving knowledge creation and sharing, competitions and awards, brokerage between innovators and entrepreneurs, and retail, mostly at the village level. These activities aim to build rural confidence and networks (World Bank 2016).

Start-ups. Vietnam promotes start-ups in agriculture as well as other areas. Mekong Business Initiative's (MBI's) Innovation Challenge promotes the incubation and acceleration of enterprises by helping start-ups access a larger pool of resources. In addition to assisting start-ups with access to finance, MBI supports mentorship programs to help start-ups develop their business management skills. The program also extracts bottom-up policy lessons by piloting innovative business models in partnership with young (and especially female) entrepreneurs. MBI has organized different acceleration programs and competitions in four sectors: the Mekong Agriculture Technology Challenge, Fintech Challenge Vietnam, Smart City Innovation Challenge, and Mekong Innovative Startup Tourism (Austrade 2019).

Grants and soft loans. Thailand's National Innovation Agency (NIA) provides grants and soft loans for innovative projects in firms primarily in the areas of bio business, eco-industry, and design solutions. The NIA has four financing schemes: technology capitalization for testing prototypes, zero-interest innovation projects to secure low-cost loans from banks for technology development, cluster-based innovation projects primarily in biotechnology fields, and venture capital to initiate production (this last program has been discontinued). The NIA has also established an Innovation Management School, which provides training for executives (Intarakumnerd 2010).

Venture capital. Venture capital for agriculture is available in China and Malaysia. In China, however, venture capital and start-ups have yet to reach the level seen in the developed world. Most of the deals fall in the expansion stages, ranging between $1 million and $50 million. Concerted investment and acquisition by the country's largest technology companies, such as Baidu, Alibaba, and Tencent, which invested at least $741 million in agri-food start-ups in 2017, is changing the scene (Burwood-Taylor 2018). Major concerns relate to having an appropriate enabling environment, that is, rules governing finance and investment, and the need to facilitate ongoing education and engagement in food system issues and opportunities in both the start-up and investor communities in China (WEF 2018).

The Malaysian Life Sciences Capital Fund for technology transfer is a public-private venture fund, focused on importing technologies that can be adapted to the national oil palm industry. Incubators transfer technology across national and corporate borders in various ways, including through intellectual property markets, manufacturing contracts, and joint ventures (World Bank 2016). Malaysia has promoted venture capital for years, but firms have made limited use of it. Restrictive investment criteria, poorly communicated business plans, low public awareness, a general disconnect between the potential entrepreneurs and the venture capital industry, and a lack of skilled personnel to manage the funds are the main obstacles for many companies (OECD 2016).

The public sector may also facilitate and provide incentives for co-innovation through different innovation funds (matching grants, competitive research grants, consortia) and other resources to initiate and sustain novel partnerships (World Bank 2012). Often such innovation funds set criteria to address market failures (for example, engagement with or services to smallholders, e-services, precision agriculture, sustainable practices in commodity production) and may create opportunities to address emerging needs.

Venture capital targeting the agri-food system is still limited in the region. The agriculture sector is far behind other sectors in venture capital investment. However, the scene began to change with the 2007–08 food price crisis, and global venture capital firms have begun showing interest in the agriculture sector. The Southeast Asian market has great potential if the solutions are for the smallholder market (Burwood-Taylor 2018; OECD 2013). Venture capital investment in agriculture has often been constrained by perceptions of high risk and low returns. Governments can foster an attractive environment for venture capital funds and corporate ventures focusing on agricultural innovation and help ensure that private sector investments make a greater impact. Box 6.11 features experiences in soft loans and venture capital in China, Malaysia, and Thailand.

Area-based approaches that combine incentives, support, and infrastructure can foster firm-level innovation. Clusters that enable stakeholders from a subsector or a value chain to benefit from economies of scale, geographic proximity, and complementary public investments have also been used in the region. China has attracted private investment particularly through its special economic zones and many other area-based instruments, such as science parks, demo zones, and high-tech parks (box 6.4). Special economic zones have contributed significantly to innovation in China, for example, through technology transfer and business model innovation. One of Thailand's key innovation policies has been promotion of the cluster approach, especially in the food sector (box F.8 in appendix F). Vietnam has also implemented area-based approaches, attracting private investment through high-tech agro parks (box 6.6) and an innovation challenge fund for start-ups (box 6.11).

Although the innovation scene is changing quickly in the region, some aspects remain the same. All the instruments require cooperation between governments, private agribusiness, and farmers, and tend to work best when the policy and regulatory environment is most appealing to the private sector.

INNOVATION CAPACITY AND SKILLS FOR LONG-TERM SUSTAINABILITY

Agricultural education and training institutions develop human resources and serve as a source of knowledge and innovations for the agri-food system. Successful transition to an innovation-driven agri-food system therefore requires consistent efforts to increase the education level and skills of farmers, firms, scientists, and AIS decision-makers. The primary constraint is that institutions have not kept pace with the labor market's demand for knowledge and practical competencies, especially in agribusiness and program management and in the problem-solving and interpersonal skills that are essential for actors to function in an AIS (World Bank 2012). ICT has also taken on a prominent position in the development of the agri-food sector (chapter 5), and its role in enhancing the functions of education institutions, curricula, and graduates' skills is warranted.

Education institutions can also play a significant part in promoting collaboration across disciplines and between the public and private sectors.

Despite the differences in size and quality of tertiary and technical and vocational education and training (TTVET) institutions across developing East Asia (table F.1 in appendix F), they share a few common TTVET-specific challenges. The TTVET institutions suffer from an increasing brain drain of young and educated graduates (in agriculture along with other disciplines) and difficulties in attracting students and lecturers to the sector. The undervalued TTVET system is inadequate for meeting labor market demand (for example, for business, management, ICT, and cross-disciplinary skills). Both tertiary and vocational curricula are mostly supply-driven and outdated. In addition, cross-disciplinary collaboration and skills are limited, which constrains discovery of solutions to emerging cross-sectoral issues. The deficiencies in the training of workers by firms, especially among small and medium enterprises, limits knowledge upgrading and undermines progress toward higher value added and productivity. In a similar vein, aging (and sometimes absent) farmers with lower education and skills add to difficulties in improving the productivity and sustainability of farming. However, several reforms (for tertiary and vocational education and for farmer training) have been carried out to address the relevance and attractiveness of the agriculture sector as well as skills shortages and multidisciplinary collaboration, although mostly in the upper MICs and the region's high-income countries. These reforms are discussed in this section. Table F.1 and appendix F provide additional details on selected approaches.

The quality, relevance, and efficiency of agricultural education constrain the development of human resources in most countries. Enrollment in agricultural programs is low or declining in some countries (Cruz et al. 2013; James, Gill, and Bates 2013; UNCTAD 2018).[23] Reasons for low enrollment include quality, relevance, and access to education as well as overall declining attractiveness of the agriculture sector. China's experience in agricultural education reform at the tertiary level serves as a good example of a gradual process of change that aimed to improve the relevance and attractiveness of agricultural education (box 6.12).

Collaboration across institutions and disciplines offers opportunities to improve the relevance of education and graduates, and to improve joint problem

BOX 6.12

Agricultural tertiary education reform in China

Agricultural education reform at the tertiary level began in China in the 1980s. The objective was to better respond to the wider reforms in the agriculture sector and in the economy (toward a market orientation).

The changes that emerged from the gradual reform, which took more than a decade (including pilots and wide stakeholder consultation), include the following: decentralization of the administrative structure along with greater decision-making responsibility, changing curricula and pedagogical approaches, changing student enrollment, aligning employment patterns more closely with labor-market expectations, changing internal administrative structures for staff recruitment, and reforming logistical systems. Although several outcomes were positive, graduate employment did not improve because of greater competition for jobs.

Source: World Bank 2012.

solving and financing. Limited interaction and exchange of personnel between universities and industry have tended to exacerbate skilled labor supply shortages in the region. However, China, Malaysia, Thailand, and to some extent Indonesia have promoted university-industry collaboration. China and Indonesia have pursued such collaboration through PPPs and diverse innovation funds (see section titled "Building private sector capacity for market-based innovation is necessary for agri-food system transformation"). CoEs[24] may also foster multidisciplinary R&D collaboration, which is often a precondition for identifying solutions to, for example, nutrition, zoonoses, food safety, and environmental challenges. For instance, in the Netherlands, the functions of agricultural R&D and that of Wageningen University were combined in 1997, resulting in the establishment of the Wageningen University and Research Center. The trigger for the merger came from the Ministry of Agriculture's desire to improve the image of agriculture (among potential students as well as others) and to develop a content-inspired international learning environment with greater attention to competencies, professional and academic education, and diverse and dynamic learning tracks (World Bank 2012). A prominent regional example of university-industry links and a CoE comes from the Thai shrimp industry (box F.8 in appendix F).[25] Additional support, such as facilitation, contracting, and IPR, is often needed to accelerate innovation (see section titled "Building private sector capacity for market-based innovation is necessary for agri-food system transformation").

Most countries in the region have paid limited attention to improving the availability of, access to, and quality of the technical track. Vocational training can play a significant role in improving the skills of extension agents, business employees, and farmers. However, based on the literature review, vocational agricultural training is either limited in scope or receives scant attention. For instance, in Cambodia and Vietnam, significant deficits of skilled technicians and workers with vocational qualifications remain, despite a desperate need for people with such skills in industry (OECD 2017b). A major issue is the perceived inferior status of vocational training as compared with university qualifications. Malaysia has tried to tackle this problem by incorporating vocational training programs into tertiary education, for example, by setting up several new university colleges to revamp the science, technology, and engineering education system. A national dual training system has also been incorporated into existing vocational education to address technical labor shortages. In Thailand, a science-based technology school has been established to increase the number of qualified vocational students. Work-integrated learning is being expanded through the Practice Engineering School approach to meet industry demand for engineers. The program provides work and research experience at industrial sites to students who study for one year and work on industrial projects for the second year at the company (Intarakumnerd 2010). Business firms are also important creators of human capital for the innovation system and are not simply employers (OECD 2013; appendix F). A few recent approaches (box 6.13) illustrate that improvement of vocational training is possible.

Improved farmer capacity, along with other investments, is a precondition for long-term development of the agri-food system. The past several decades have seen significant advances in average schooling levels of the labor force across the developing world. However, average attainment levels and the quality of rural schooling trail that of urban areas (Barrow and Lee 2013). Although adult literacy rates are high in most East Asian countries, the differences

BOX 6.13

Approaches to addressing the availability of, access to, and quality of vocational schools

Singapore has been particularly effective in promoting workforce development through a network of institutes for technical education with a host of industrially relevant vocational training programs. Many of these programs are collaborative ventures between the government and reputable overseas partners (OECD 2013).

The *Republic of Korea* has also invested in vocational training in agriculture; in 1997, the government established the Korea National College of Agriculture and Fisheries as a professional school to foster future leaders in agriculture. Students are exempt from tuition and admission fees for three-year programs, and the government supports other expenses. Graduates are eligible for a subsidy for young farmers but are required to engage in farming for at least six years (OECD 2018b). For further details on Korea, see appendix E.

In *Australia,* AgriFood Skills Australia was established in 2004 to provide accurate industry intelligence on current and future skills needs and training requirements for the agri-food industry. Since then the training packages for vocational training have been regularly updated to meet sector needs (World Bank 2012).

between countries in secondary education are wide, suggesting significant variation between farmer education levels across the countries (table F.2 in appendix F). Many countries are also experiencing a substantial learning gap (appendix F) (World Bank 2019). Even though human capital is a central driver of sustainable growth and poverty reduction, policy makers often find it hard to make the case for human capital investment, particularly in the human capital of young children who will not join the workforce immediately (World Bank 2018b).

Several countries also face a new challenge—aging farmers, who tend to have fewer years of education and more limited skills. This deficit can have significant implications for farmers' ability to adopt technically challenging innovations. China and Thailand have explored ways to improve farmers' capacity to meet the requirements of modern agriculture and new innovative tools (appendix F). As discussed in this chapter's earlier section titled "Misalignment between sector-wide vision and innovation capacity, investment, and governance," extension programs may be suitable for the training of farmers in most contexts, but the need for technical track and college education is expected to rise in increasingly urban contexts.

BETTER RESOURCE USE AND INNOVATION OUTCOMES FROM STRONGER GOVERNANCE OF AIS

The governance of AIS is expected to ensure that national priorities are coordinated and communicated clearly, that progress is monitored, and that policy outcomes and impacts are evaluated against defined objectives. The integration of AIS into overall governance of the NIS ensures better use of public funds (by creating synergies and avoiding unnecessary duplication) and increased efficiency through the pooling of different types of expertise. AIS investments must also be specific to the context and respond to the stage of development in each country and agriculture sector. The scale of operations is also likely to vary from local or zonal to subsectoral or national. This variation requires investments to be assessed, prioritized, sequenced, and tailored to the needs, challenges, and resources that are present (World Bank 2012).

Ideally, a nationally mandated but independently governed agricultural innovation council or committee would coordinate the development of a strategic vision for agricultural innovation. This body may also coordinate and formulate the corresponding agricultural innovation policy (to be increasingly integrated into general STI policy), design agricultural innovation priorities and agendas, and monitor and evaluate innovation programs and their impact. In theory, responsibilities for policy making, financing, and implementation should be separate, but experience varies in practice. Many innovation councils are advisory and policy-making bodies with no mandate to channel funds; others have been more effective at inducing coordination of policy when they control innovation funds (box 6.2; World Bank 2012).

STI councils are used as a mechanism to respond to the growing need for more effective innovation governance in some East Asian countries. These councils can fulfill the functions of priority setting and advising, policy coordination, and strategic planning. In a recent benchmarking study, most of the councils reviewed (Austria, Canada, Denmark, Germany, Singapore, Switzerland, the United Kingdom, and the United States) were found to primarily serve an advisory function. Some councils also have a coordination function (China, Finland, Germany, Japan, and Korea), a priority-setting function (China, Finland, Japan, and Korea), or even a policy planning function (Finland). Their role in evaluation tends to be limited, but mostly because of a lack of institutionalization of evaluation rather than for reasons of good design. International experience also shows that the coordination function is the most difficult to achieve (OECD 2016). The National Innovation Agency of Thailand (box 6.14) serves as an example from the region of a multisector coordination agency.

The governance arrangements for AIS vary across developing East Asia. In all countries, the line ministry for the agricultural sector, or a specialized institution or agency under it, performs oversight functions for AIS. However, the level of integration between AIS and the NIS varies. In Indonesia, Malaysia, the Philippines, Thailand, and Vietnam, agriculture is to a greater or lesser extent integrated into NIS frameworks. In these countries, other ministries, along with agencies in the STI policy space, also have a role in the governance of AIS through their broader policy formulating, coordinating, and implementing roles (OECD 2017b). In Cambodia and Lao PDR, the ministries of agriculture head the AIS, and are the main performers of agricultural research through the research institutes that operate under their responsibility. In Malaysia and Myanmar, several ministries have responsibilities for agricultural sector policy and research. The AIS governance structure and main coordination mechanisms and knowledge brokers engaged in AIS are summarized in table F.3 in appendix F.

The overall institutional structure of R&D and AIS affects the efficiency of AIS governance. Greater attention to strong R&D coordination at the national level is warranted to improve the overall efficiency of AIS governance. In general, most agricultural researchers are employed in government research agencies, sometimes in a relatively small number of research agencies, as in China, Indonesia, and Malaysia (section titled "Increasing agricultural R&D financing to support developing East Asia's response to emerging needs"). In other

BOX 6.14

Thailand's National Innovation Agency

Thailand's National Innovation Agency (NIA), established in 2003, supports the development of innovations to enhance national competitiveness and pays significant attention to agriculture and other biological sciences. The NIA operates under the overall policy guidance of the Ministry of Science and Technology. In 2010 the NIA was upgraded by government decree from a project within the Ministry of Science and Technology to a public organization. Its new board comprises representatives from key government agencies as well as the private sector, and it is chaired by the executive chairman of Bangkok Bank.

The NIA is unusual in that it offers direct financial support to private companies for innovation-related projects. In 2009, it supported 98 innovation projects initiated by private companies. The agency essentially shares the investment risks associated with innovative, knowledge-driven businesses through technical and financial mechanisms. NIA's main strategies are the following:

- *Upgrade innovation capability*, with a focus on bio-business, energy and environment, and design and branding. The NIA encourages the development of start-ups and supports commercialization of research.
- *Promote innovation culture* within organizations of all types. The NIA operates an innovation management course for executives, the National Innovation Awards, an innovation ambassador scheme, an Innovation Acquisition Service, and a Technology Licensing Office.
- *Build up the national innovation system*. Although the NIA is in an ideal position to propose measures to enhance policy coherence across ministries, its portfolio suggests that it focuses more on discrete and disguised subsidies for firm-level innovation.

One challenge is that the NIA's definition of its role as "coordinating industrial clusters both at policy and operational levels, promoting innovation culture, and building up innovation systems, with a broader aim to transform Thailand into an innovation-driven economy" appears to overlap with the mandates of the recently established Office of Science, Technology, and Innovation; the National Economic and Social Development Board; the National Science and Technology Development Agency; and the Office for Small and Medium Enterprise Promotion.

Sources: Brimble and Doner 2007; Wyn Ellis, personal communication; Intarakumnerd, Chairatana, and Tangchitpiboon 2002; World Bank 2012; NIA, www.nia.or.th.

countries, the structure of public agricultural research systems is more complex, with multiple government levels and HEIs engaged in agricultural R&D (Cambodia, Lao PDR, Myanmar, the Philippines, Thailand, Vietnam), often contributing to inefficiencies. Such inefficiencies may be mitigated by strong coordination across different agencies, as attempted in Vietnam (through the Vietnam Academy of Agricultural Sciences) and the Philippines (by the Philippine Council for Agriculture, Aquatic, and Natural Resources R&D). However, the institutions focus primarily on research, not on wider innovation, which may limit their effectiveness in influencing wider AIS and NIS policies (box 6.2).

Knowledge brokers, coordinators, and platforms have the potential to play an important formal and informal role in coordination and innovation processes. They can ensure flows of knowledge, capital, information, and skilled personnel, as well as provide inputs to innovation policy development, plans, and implementation (World Bank 2012) (box F.4 in appendix F). Innovation brokering and platforms are particularly well-suited to strategic tasks such as response to climate change or zoonoses. In China, industry associations and farmer cooperatives function as a bridge between farmers and other AIS actors. In Indonesia, Malaysia, and Thailand, commodity boards and associations serve as nodes between public and private actors, whereas in Vietnam some public agencies and incubators serve as brokers. Overall, countries still have limited experience in facilitating networks and platforms for agricultural innovation (see box 6.3 on Chile).

Assessment of innovation policies and instruments as part of countries' decision-making processes and governance is not mainstreamed in developing East Asia. The rationale for investing in assessment, priority setting, and monitoring and evaluation within an AIS is that practitioners require information for short- and long-term decision-making and for managing limited resources effectively within complex processes of technical and institutional change that seek to achieve social, economic, and environmental goals. Assessments are needed to inform decision-making at policy, investment, organization, and intervention levels (World Bank 2012). Increasing coordination between innovation and research institutions and strengthening evaluation mechanisms would also improve the efficiency of public funding to agricultural R&D. However, limited data availability on, for example, monitoring and evaluation, other assessments, and foresighting suggests that these methods are not yet part and parcel of innovation policy governance (table F.3 in appendix F). Lack of institutionalization of evaluation has also been found to be common in many OECD countries (OECD 2016).

IMPORTANCE OF THE ENABLING ENVIRONMENT TO INNOVATION

Innovation policy is not likely to compensate for seriously flawed framework conditions. In many instances, innovation and business development do not occur without complementary investments in infrastructure, IPR protection, regulation, finance, and other factors such as land rights to create a supportive environment. Table 6.4 summarizes the countries' key enabling-environment issues.[26]

IPR are an important factor influencing the performance of AIS. Adequate protection of IPR enhances private R&D investment, including FDI, in agriculture (World Bank 2012). Innovation in plant varieties tends to be cumulative, that is, prior knowledge is needed to come up with new innovations. The adoption of new plant varieties also depends on efforts to adapt innovation developed elsewhere to breed locally suitable varieties (Evenson and Gollin 2003). These characteristics may make IPR associated with new plant varieties a critical issue.

Although most countries in the region have IPR regulation and laws (including for plant variety protection) in place, differences exist regarding their approval of the International Convention for the Protection of New Varieties of Plants (UPOV) (the 1991 Act) (table 6.4). All countries have joined the World IPR Organization and the World Trade Organization Agreement on Trade-Related Aspects of Intellectual Property Rights. They also have their own patent laws in place[27] (OECD 2017b). China and Vietnam are the only countries of the ones under consideration that have joined the UPOV. However, Malaysia and the Philippines have initiated the procedure for acceding to the UPOV Convention,

TABLE 6.4 **Enabling environment for agricultural innovation in select countries**

COUNTRY	INTELLECTUAL PROPERTY RIGHTS	REGULATORY FRAMEWORKS	INFRASTRUCTURE: ROADS, IRRIGATION, ENERGY, MARKET, ICT CONNECTIVITY	FINANCE	LAND
China	WIPO (2007) WTO-TRIPS (2001) Patents (1994) Copyright (1992) Trademarks Plant Variety Protection (1999) UPOV Convention member (1999) IPR are not properly protected, serving as a barrier to FDI and domestic private R&D	The Food and Drug Administration was established in 2013. The Food Safety Law was revised in 2015 to establish a more scientific and strict supervision system for food safety. Biosafety regulation on transgenic organisms updated in 2016; plan to establish biosafety evaluation system, but enforcement shows weaknesses. Regulatory framework for environmental protection in place but poorly enforced. Regulation on pesticide use strengthened.	Good road and rail infrastructure—higher than OECD average. ICT connectivity: electricity in 100% of the villages. ICT connectivity significantly improved but still limits widespread use of e-services and precision agriculture.	Affordability and availability of financial services are limited. Tax incentives for private sector R&D in place. Barriers to FDI: foreign companies are not allowed to conduct transgenic crop breeding. Nontransgenic plant breeding and seed production are "restricted" and require foreign investors to establish a joint venture with Chinese firms.	Land reform instituted enabling consolidation, sale, and rent as well as use as collateral.
Indonesia	WIPO (1979) WTO-TRIPS (1995) Patent (2001) Copyright (2014) Trademarks (2001) Plant Variety Protection (2000) Weak enforcement of patents and licenses a disincentive to private sector	Relatively weak food safety regulations and enforcement. Biosafety regulations in place, enforcement shows weaknesses.	Infrastructure requires strengthening albeit irrigation infrastructure has improved. ICT connectivity limiting widespread use of e-services and precision agriculture.	Farmer access to finance is limited.	Land availability and land market rights require strengthening.
Malaysia	WIPO (1989) WTO-TRIPS (1995) Patent (2006) Copyright (2006) Trademarks (2002) Plant Variety Protection (2004)	Food safety and quality standards in place. Enforcement of environmental regulations good. Biosafety regulations in place.	Good infrastructure and ICT connectivity.	Access to finance is relatively good. Tax incentives for private sector investment are available.	Land market rights and access require strengthening.
Philippines	WIPO (1980) WTO-TRIPS (1995) Patent (1998) Copyright (2013) Trademarks (1998) Plant Variety Protection (2002)	Relatively weak food safety regulations and enforcement. Environmental regulations and their enforcement are relatively weak. Biosafety regulations in place.	Infrastructure, including in agriculture, requires strengthening. ICT connectivity limits widespread use of e-services and precision agriculture.	Farmers' access to finance is limited. Tax incentives for private sector R&D in place.	A complex issue—current land reform redistributes private land; land leasehold system in place.

(continued)

TABLE 6.4, *continued*

COUNTRY	INTELLECTUAL PROPERTY RIGHTS	REGULATORY FRAMEWORKS	INFRASTRUCTURE: ROADS, IRRIGATION, ENERGY, MARKET, ICT CONNECTIVITY	FINANCE	LAND
Thailand	WIPO (1989) WTO-TRIPS (1995) Patent (1999) Copyright (2015) Trademarks (2000) Plant Variety Protection (1999) Plant Breeder's Rights Regime enforced IP support from science parks	Food safety and quality standards in place. Enforcement of environmental regulations. No biosafety regime; significant problem for biotechnology innovation. A draft submitted to WTO for review.	Infrastructure and well-functioning markets. ICT connectivity limits widespread use of e-services and precision agriculture.	Farmers' access to finance is relatively good. Tax incentives for private sector R&D in place.	Strong land market rights.
Vietnam	WIPO (1976) WTO-TRIPS (2007) Patent (2009) Copyright (2009) Trademarks (2009) Plant Variety Protection (2004) UPOV Convention member (2006) Weak IPR protection and market for technology, disincentive for researchers and private sector	Relatively weak food safety regulations and enforcement. Biosafety regulations in place, enforcement shows weaknesses.	ICT connectivity limits widespread use of e-services and precision agriculture.	Access to finance a major constraint for private sector. New incentives have been put in place to support start-ups in agri-food sector. Tax incentives for private sector R&D in place.	Land rental enables farm consolidation and mechanization but is not consistently practiced across the country.

Sources: Original table for this publication. FAO GM Platform 2020 (http://www.fao.org/food/food-safety-quality/gm-foods-platform); OECD 2016, 2017a, 2017b, 2018a; UNCTAD 2018; World Bank, n.d., 2016, 2019. Note: FDI = foreign direct investment; ICT = information and communication technology; IPR = intellectual property rights; OECD = Organisation for Economic Co-operation and Development; R&D = research and development; WIPO = World IPR Organization; WTO-TRIPS = World Trade Organization-Trade-Related Aspects of Intellectual Property Rights.

and Cambodia, Indonesia, Lao PDR, Myanmar, and Thailand have been in contact with the UPOV Office for assistance in developing laws based on the UPOV Convention (OECD 2017b; UPOV 2020). Cambodia, Lao PDR, and Myanmar do not yet have laws governing the protection of plant varieties.

IPR legislation in developing East Asia is generally adequate, but the institutional capacity to manage and enforce IPR is weak in some countries. Although IPR policies and regulations are largely in line with international rules and guidelines, China's protection of IPR still lags behind that of most OECD countries, particularly with regard to enforcement (OECD 2016, 2018a). In China, which is at the forefront in the use of GM and gene editing technology globally, lack of IPR protection has slowed both public and private R&D and commercialization of GM crops (box 5.5; OECD 2018a). The developing East Asian countries other than China, Malaysia, and Thailand often lack the institutional capacity to manage and provide legal support to IPR cases. For instance, in Vietnam, weak patent and plant variety protection limits the performance of its AIS. In Indonesia, weak enforcement of IPR, such as for patents and licenses, is a disincentive for private sector investment in agricultural R&D (OECD 2017b).

Regulations provide a basic economic environment within which all firms operate and make investment decisions. Regulations also play an important role in the creation of incentives to use natural resources in an environment-friendly manner. Although standards are an important element of regulations, only food safety and biosafety are addressed here because of data limitations.

All developing East Asian countries have food safety regulations in place; however, their capacity to respond to new challenges and to enforce the regulations varies significantly (table 6.4 and the section titled "Readiness of developing East Asian countries to embrace transformative innovations" in chapter 5). The region's high-income countries, such as Korea and Singapore, have well-established food safety systems (World Bank 2019). Whereas China, Malaysia, and Thailand are building their capacity and modernizing their food safety systems, Indonesia, the Philippines, and Vietnam have significant challenges in meeting the food safety requirements of their quickly changing agri-food systems (World Bank 2019). Cambodia, Lao PDR, and Myanmar[28] have limited capacity; however, their food safety concerns are not yet as prominent.

A well-defined and consistently applied biosafety regulatory system can be a powerful stimulus for investments in innovation. Investments needed to operationalize a biosafety regulatory system should promote interministerial cooperation, sound and pragmatic policy development, scientific risk assessment and risk management, rational inspection and enforcement activities, and meaningful stakeholder consultation and public participation (World Bank 2012). Most developing East Asian countries have devised or are in the process of devising regulatory systems to manage biosafety and enable innovation in GM products (tables 5.3 and 6.4), yet enforcement of biosafety rules has been challenging (in, for example, China, Indonesia, Vietnam). Although Thailand does not have a biosafety regime in place, it has submitted a draft for World Trade Organization review. Lao PDR has no biosafety law.[29]

Efficient and well-developed infrastructure plays an important role in connecting farms and firms to market opportunities, innovations, and services. In contrast, inefficient and underdeveloped infrastructure can significantly increase farms' and firms' costs (OECD 2017b). The countries' infrastructure (roads, railway, irrigation, electricity, market, ICT connectivity) varies significantly (table 6.4). China has drastically improved its transport infrastructure, with almost all administrative villages having gained access to the public road system by the end of 2006 (OECD 2009). China has developed an overall higher-quality transport infrastructure than the OECD average (OECD 2018a). Although ICT connectivity in rural areas is still incomplete, China has higher ICT penetration than the other BRICS countries (Brazil, Russia, India, and South Africa) but lower penetration than the OECD average (OECD 2018a). Similarly, Malaysia and Thailand have good infrastructure, although ICT connectivity is still lacking in Thailand (OECD 2017b). In the mid-range and lower MICs, inadequate infrastructure investment still limits private sector potential.

Access to finance by farms and firms has remained stubbornly difficult in the agri-food sector (table 6.4). Other than in Malaysia and Thailand, financial services to farmers and other private actors are an investment-limiting factor in developing East Asia (OECD 2017b, 2018a) (also see the sections titled "Providing the incentives and breaking the barriers to increase private sector R&D" and "International collaboration for a regionwide response to agri-food system challenges"). In Vietnam, for instance, limited access to finance is a significant constraint for private sector investment, despite the recent introduction of tax incentives for start-ups. In China, affordability and availability of finance services is limited. In addition, foreign investment in agricultural innovation is restricted, for example, on transgenic crops (OECD 2016, 2018a).

Land rights and access can have significant impacts on farmer and firm adoption of innovations. Secure land market rights and access support sustainable agricultural productivity growth by enhancing landholders' incentives to make long-term investments (World Bank 2012). Moreover, secure land market rights facilitate the use of land as collateral for loans, which can be important for funding productivity- and sustainability-enhancing innovations. Many developing East Asian countries (for example, China, the Philippines, Vietnam) have carried out consecutive land reforms that enable the use of land as collateral and enable land consolidation, rental, and sale. However, in many countries (Indonesia, Malaysia, the Philippines) land rights still impede investment (table 6.4). For example, in Malaysia strengthening land market rights and access may enhance the capacity of Malaysia's farmers and firms to increase agricultural productivity (OECD 2017b). In Indonesia, complex and insecure land rights have impeded firms' interest in investing in agriculture (FAO, n.d.).

NOTES

1. For instance, Malaysia places significant emphasis on agriculture and innovation as drivers of productivity, sustainability, inclusivity, and growth (National Agri-Food Policy 2.0) and emphasizes the role of new agricultural technology (information and communication technology, precision agriculture).
2. China's net FDI to agriculture (targeting Africa and Latin America) increased from $190 million in 2006 to $3.29 billion in 2016, an average annual increase of 33 percent (Jiang, Chen, and Wang 2018). Net FDI has resulted in, for example, land acquisitions and imports of grains by state-owned enterprises (Zambon et al. 2019).
3. China, Malaysia, Thailand, Indonesia, the Philippines, Vietnam, Cambodia, Lao PDR, and Myanmar.
4. Examples of such plans include China's 13th Five-Year Plan (2016–2020), which called for an innovation-driven development strategy; Malaysia's Agrofood Policy 2.0, 10th Malaysia Plan (2011–15), and Economic Transformation Programme; Thailand's Agriculture 4.0, 12th National Economic and Social Development Plan (2017–2021), and National Science Technology and Innovation Policy and Plan; Indonesia's Smart Farming 4.0 and Master Plan for the Acceleration and Expansion of Economic Development; Vietnam's Master plan on high-tech agricultural parks and zones through 2020, with a vision toward 2030 and The National Assembly Plan; the Philippines' Smarter Agriculture and The National Science and Technology Plan (2002–20).
5. Across 21 European Union and OECD countries, the share of farmers with no formal training in agriculture was 73.8 percent in 2005 and 55 percent in 2010, with basic training was 14 percent in 2005 and 34.5 percent in 2010, and with full training was 12.2 percent in 2005 and 10.4 percent in 2010 , based on data from EUROSTAT's Farm Structure Survey (https://ec.europa.eu/eurostat/statistics-explained/index.php?title=Farm_structure_survey_-_survey_coverage)
6. Korea has tried to improve both the attractiveness of agriculture among youth and farmer education levels. The share of agricultural high school graduates who become self-employed farmers is about 1 percent. At the tertiary level of education, approximately 30,000 students are enrolled in agricultural colleges and universities; of these, the rate of becoming a self-managed farmer is about 7 percent (OECD 2018b).
7. The focus of investment varies widely from economy to economy.
8. For example, the Netherlands has moved from potatoes to seed potatoes and from protected horticulture production to a strong global position in the high-tech industry of greenhouse construction (Dons and Bino 2008).
9. China spent about 2.1 percent of GDP on overall R&D in 2015, which is close to the EU15 (Austria, Belgium, Denmark, Finland, France, Germany, Greece, Ireland, Italy, Luxembourg, the Netherlands, Portugal, Spain, Sweden, United Kingdom) average, and slightly lower than the OECD average of 2.4 percent. The gap with the OECD average has been narrowing compared with the early 2000s (OECD 2016).

10. The Indonesian Agency for Agricultural R&D oversees nine major research centers on crop and livestock R&D and is linked with the Indonesian Research Institute for Estate Crops. The Forest R&D Agency oversees most of the country's forestry R&D. The higher education sector plays a fairly important role as well.
11. Private sector engagement is also minimal; for instance, in Myanmar, only one company carries out R&D other than some outsourced R&D (outside Myanmar) (Stads and Kam 2007). Cambodia has limited R&D in industrial crops (OECD 2017b).
12. For example, China restricts R&D into GM crops; a majority of public R&D is directed to experimental development.
13. The market failure argument was originally formulated to bolster public support for basic research, but an economic study finds that large gaps in the social and private rates of return apply to a wide range of industries, including agriculture (Pardey, Alston, and Ruttan 2010).
14. Private sector activity is diverse; for example, in value chains not all activity is related to innovation. This review focuses primarily on innovation.
15. For instance, weak enforcement of IPR; restrictions on private sector plant breeding and GM crops development; heavy allocation (three-quarters of public sector focus) toward experimental development that is typically considered to be a private sector niche (OECD 2016).
16. Private R&D is focused on hybrid seed, genetics of rubber trees, cultural practices, post-harvest, livestock industry (feed, breed), and crop protection.
17. Consortia are research-innovation alliances that typically involve competitive and matching funding and diverse stakeholders, and address the entire innovation process (from identification to adoption). Chile and India, for instance, have significant experience in consortia. Other instruments for co-innovation are discussed in the section titled "Building private sector capacity for market-based innovation in the agri-food system."
18. The Philippines views its partnership with the International Rice Research Institute as an investment to help address national and global agriculture. The Philippines' partnerships with the CGIAR centers have grown and expanded from joint research undertakings to institutional development. The benefits include institution building, technology development and varietal improvement, training and technology sharing, and research capability building (OECD 2017b).
19. Mobile network operators, for example, can invest by providing large text packages at a lower price, collecting premiums, distributing payments, or participating in extending networks to rural areas.
20. For instance, Indonesia aspires to encourage 1 million farmers and fishermen to use Go Online by 2020 and grow 1,000 tech start-ups by 2020. Thailand pursues, for example, Smart Farmer e-learning to improve farmers' and extension agents' capacity (UNCTAD 2018).
21. Based on data from the Global Forum for Rural Advisory Services World Wide Extension Study data set (https://www.g-fras.org/en/world-wide-extension-study.html).
22. Such measures may raise awareness of networking opportunities and help search for partners, organize financing and operating networks, provide an interface for scientific and innovation networks through PPPs, and create international links and build global networks (OECD 2013).
23. In the Philippines, enrollment in agriculture, forestry, and fisheries programs has declined (Cruz et al. 2013). In Thailand, universities' agriculture departments struggle to recruit students and lecturers (box F.7 in appendix F) (UNCTAD 2018). In Cambodia (2009–10), students enrolled in agriculture and in animal science and veterinary medicine accounted for 2.3 percent and 0.5 percent (of all students enrolled), respectively (James, Gill, and Bates 2013).
24. CoEs are a policy measure used by many governments to promote a robust R&D and innovation environment by encouraging institutional profiling and generating a critical mass of researchers. CoEs are often located in HEIs (Hellstrom, n.d.).
25. China, Malaysia, and Thailand have all explored different forms of CoEs agricultural science (and technology) parks in China (CDB 2015; OECD 2018a), biotechnology CoEs (among others) in Malaysia (OECD 2016), and several science parks or CoEs in Thailand (UNCTAD 2018).
26. Note that data limitations on the enabling environment may limit the findings in this section.
27. Myanmar's legislation may not be compliant with the minimum provisions of the TRIPS Agreement (OECD 2017b).

28. For example, in Myanmar, the Food and Drug Administration under the Ministry of Health is the regulatory body. Enforcement is weak given multiple ministerial roles in food safety matters.
29. Based on data from the FAO GM Foods Platform, Food and Agriculture Organization of the United Nations (http://www.fao.org/food/food-safety-quality/gm-foods-platform).

REFERENCES

ADB (Asian Development Bank). 2019. *Policies to Support Investment Requirements of Indonesia's Food and Agricultural Development during 2020–2045*. Metro Manila, Philippines.

Alston, J. M. 2010. "The Benefits from Agricultural Research and Development, Innovation, and Productivity Growth." OECD Food, Agriculture and Fisheries Working Paper 31, OECD Publishing, Paris.

Alston, J. M., C. Chan-Kang, M. Marra, P. G. Pardey, and T. Wyatt. 2000. "A Meta-Analysis of Rates of Returns to Agricultural R&D." International Food Policy Research Institute, Washington, DC. http://www.ifpri.org/sites/default/files/publications/rr113.pdf.

Aquino, A. P., A. B. D. Correa, J. R. A. Manalo, and P. S. Faylon. 2014. "Public Sector Investments in Science, Technology and Innovation for Inclusive Growth and Competitive Economy in the Philippines: A Focus on Agriculture." Food and Fertilizer Technology Center Agricultural Policy Articles, Taipei. http://ap.fftc.agnet.org/ap_db.php?id=215&print=1.

Austrade. 2019. "Vietnam's Innovation Ecosystem." Commonwealth of Australia, Austrade, Hanoi and Ho Chi Minh City.

Barrow, R. J., and J. W. Lee. 2013. "A New Data Set of Educational Attainment in the World, 1950–2010." *Journal of Development Economics* 104: 184–98.

Bientema, N., G. Stads, K. Fuglie, and P. Heisey. 2012. "ASTI Global Assessment of Agricultural R&D Spending: Developing Countries Accelerate Investment." International Food Policy Research Institute, Washington, DC.

Brimble, P., and R. F. Doner. 2007. "University-Industry Linkages and Economic Development: The Case of Thailand." *World Development* 35 (6): 1021036.

Burwood-Taylor, Louisa. 2018. "China's Tech Giants Drive $1.8bn Investment in Chinese AgriFood Startups—New Report!" Agfunder Network Partners, September 25, 2018.

Chandra, Vandana, ed. 2006. Technology, Adaptation, and Exports: How Some Developing Countries Got It Right. Washington, DC: World Bank.

CDB (China Development Bank). 2015. "China's Special Economic Zones. Experience Gained." World Bank—Investing in Africa Forum, Addis Ababa, June 30–July 1.

Chen, C. 2018. "The Liberalisation of FDI Policies and the Impacts of FDI on China's Economic Development." In *China's 40 Years of Reform and Development: 1978–2018*. Canberra, Australia: ANU Press.

Chen, K. Z., K. Flaherty, and Y. Zhang. 2012. "China: Recent Developments in Agricultural Research." ASTI Country Note, International Food Policy Research Institute, Washington, DC. https://www.asti.cgiar.org/pdf/China-Note.pdf.

China, Ministry of Science and Technology. 2016. *Technical Specification of Evaluation for Agricultural Science and Technology Achievements*, GB/T 32225-2015 (in Chinese).

Christoplos, I. 2010. *Mobilizing the Potential of Rural and Agricultural Extension*. Rome: Food and Agriculture Organization of the United Nations (FAO) and the Global Forum for Rural Advisory Services (GFRAS).

Cruz, R. V. O., R. B. Bantayan, L. D. Landicho, and N. C. Bantayan. 2013. "Reformulating Agriculture and Forestry Education in the Philippines: Issues and Concerns." *Journal of Developments in Sustainable Agriculture* 8: 49–62.

Davis, K., B. Swanson, D. Amudavi, D. A. Mekonnen, A. Flohrs, J. Riese, C. Lamb, and E. Zerfu. 2010. "In-Depth Assessment of the Public Agricultural Extension System in Ethiopia and Recommendations for Improvement." IFPRI Discussion Paper 01041, International Food Policy Research Institute, Washington, DC.

Dercon, S., D. O. Gilligan, J. Hoddinot, and T. Woldehanna. 2008. "The Impact of Agricultural Extension and Roads on Poverty and Consumption Growth in Fifteen Ethiopian Villages." IFPRI Discussion Paper, International Food Policy Research Institute, Washington, DC.

Diem, H. X., and D. T. T. Thuy. 2019. "Investment in Agriculture in Recent Times: The Case of Vietnam." In *New Trends and Challenges for Agriculture in the Mekong Region: From Food Security to Development of Agri-Businesses*, edited by Shozo Sakata. BRC Research Report 25. Chiba, Japan: Bangkok Research Center.

Dons, H. J. M., and R. J. Bino. 2008. "Innovation and Knowledge Transfer in the Dutch Horticultural System." In *Pathways to High-Tech Valleys and Research Triangles: Innovative Entrepreneurship, Knowledge Transfer and Cluster Formation in Europe and the United States*, edited by W. Hulsink and H. Dons, 119–37. Dordrecht, the Netherlands: Springer.

EBRD (European Bank for Reconstruction and Development). 2018. "Eight Things You Should Know about Middle-Income Transitions." European Bank for Reconstruction and Development, London.

Ernst & Young. 2013. *2013 Asia Pacific R&D Incentives*. Ernst & Young Australia.

Evenson, R. E., and D. Gollin, eds. 2003. *Crop Variety Improvement and Its Effect on Productivity: The Impact of International Agricultural Research*. Wallingford, UK: CABI.

Fabregas, R., M. Kremer, and F. Schilbach. 2019. "Realizing the Potential of Digital Development: The Case of Agricultural Advice." *Science* 366 (6471). doi:10.1126/science.aay3038.

FAO (Food and Agriculture Organization of the United Nations). n.d. "Corporate Private Sector investment in Agriculture in Indonesia. A Case Study of Indonesia." FAO, Rome.

FAO (Food and Agriculture Organization of the United Nations). 2005. "Policy Issues and Investment Options to Avert Hunger and Food Insecurity in Asia. China's Agricultural Research System and Reforms: Challenges and Implications for Developing Countries." RAP Publication 2005/1. FAO, Regional Office for Asia and the Pacific, Bangkok.

FAO (Food and Agriculture Organization of the United Nations). 2006. "China's Rapid Economic Growth and Its Implications for Agriculture and Food Security in China and the Rest of the World." In *Rapid Growth of Selected Asian Economies: Lessons and Implications for Agriculture and Food Security*, edited by Jikun Huang, Jun Yang, and Scott Rozelle. Bangkok: FAO, Regional Office for Asia and the Pacific. http://www.fao.org/3/ag088e/AG088E03.htm.

FAO (Food and Agriculture Organization of the United Nations). 2009. *Rethinking Public Policy in Agriculture: Lessons from Distant and Recent History*. Policy Assistance Series 7. Rome: FAO.

Flaherty, Kathleen, Gert-Jan Stads, and Attaluri Srinivasacharyulu. 2013. *Benchmarking Agricultural Research Indicators across Asia–Pacific*. ASTI Synthesis Report. Washington, DC: International Food Policy Research Institute.

Flaherty, Kathleen, and Rozhan Abu Dardak. 2013. "Malaysia: Recent Developments in Agricultural Research." International Food Policy Research Institute, Washington, DC.

Foster, A., and M. R. Rosenzweig. 2010. "Microeconomics of Technology Adoption." *Annual Review of Economics* 2 (1): 395–424.

Fuglie, K. 2016. "The Growing Role of the Private Sector in Agricultural Research and Development World-Wide." *Global Food Security* 10 (2016): 29–38.

Fuglie, K., M. Gautam, A. Goyal, and W. F. Maloney. 2020. *Harvesting Prosperity: Technology and Productivity Growth in Agriculture*. Washington, DC: World Bank.

Group of 20. 2012. "Sustainable Agricultural Productivity Growth and Bridging the Gap for Small Family Farms." Interagency Report to the Mexican G20 Presidency.

Hall, A., L. Williams, J. Tanaya, M. Usmani, K. Esti, H. Nufaisa, and L. Xie. 2016. "ARISA: Innovation Systems Research Status and Options. Applied Research on Innovation Systems in Agriculture (ARISA) in Indonesia." Technical Report, CSIRO, Canberra, Australia.

Hall, B., and J. van Reenen. 2000. "How Effective Are Fiscal Incentives for R&D? A Review of the Evidence." *Research Policy* 29 (4–5): 449–69.

Hellstrom, Tomas. n.d. "Centers of Excellence as a Tool for Capacity Building." Synthesis Report. Programme on Innovation, Higher Education and Research for Development (IHERD). OECD, Paris. https://www.oecd.org/sti/Draft_OECD%20synthesis%20report_final.pdf.

Hoang Van Tuyen. 2017. "System of Science, Technology and Innovation in the Developing Country: The Case of Vietnam." *Vietnam Science and Technology Journal*: 4–9.

Huang, J., and S. Rozelle. 2018. "China's 40 Years of Agricultural Development and Reform." Australian National University, Canberra, Australia.

Hynes, Casey. 2018. "Vietnam May Soon Emerge as One of Southeast Asia's Blockchain Innovation Hubs." Forbes, March, 27, 2018. https://www.forbes.com/sites/chynes/2018/03/27/why-vietnam-could-emerge-as-a-blockchain-innovation-hub/?sh=6c9be1412cc5.

IFPRI (International Food Policy Research Institute). 2019. "Policies to Support Investment Requirements of Indonesia's Food and Agriculture Development during 2020–2045." International Food Policy Research Institute, Kementerian PPN/Bappenas, and Asian Development Bank, Manila.

Intarakumnerd, P. 2010. "Country Profile of Thailand for OECD Review of Innovation in South-East Asia." College of Innovation, Thammasat University, Thailand.

Intarakumnerd, P., P. Chairatana, and T. Tangchitpiboon. 2002. "National Innovation System in Less Successful Developing Countries: The Case of Thailand." *Research Policy* 31 (8–9): 1445–57.

James, D., T. Gill, and R. Bates. 2013. "Cambodia: Desktop Study." Innovation for Agricultural Training and Education, Pennsylvania State University, State College, PA.

Jiang, X., Y. Chen, and L. Wang. 2018. "Can China's Agricultural FDI in Developing Countries Achieve a Win-Win Goal? Enlightenment from the Literature." *Sustainability* 11 (1): 1–22.

Labarthe, P., and C. E. Laurent. 2009. "Transformations of Agricultural Extension Services in the EU: Towards a Lack of Adequate Knowledge for Small-Scale Farms." European Association of Agricultural Economists, 111th Seminar, Canterbury, UK, June 26–27.

Mogues, T., B. Yu, S. Fan, and L. McBride. 2012. "The Impacts of Public Investment in and for Agriculture: Synthesis of the Existing Evidence." IFPRI Discussion Paper 1217, International Food Policy Research Institute, Washington, DC.

OECD (Organisation for Economic Co-operation and Development). 2008. *Review of Agricultural Policies: Chile 2008*. Paris: OECD Publishing.

OECD (Organisation for Economic Co-operation and Development). 2009. *OECD Rural Policy Reviews: China 2009*. Paris: OECD Publishing. http://dx.doiorg/10.1787/9789264059573-en.

OECD (Organisation for Economic Co-operation and Development). 2013. *Innovation in South East Asia*. Paris: OECD Publishing.

OECD (Organisation for Economic Co-operation and Development). 2016. *OECD Reviews of Innovation Policy: Malaysia 2016*. Paris: OECD Publishing. http://dx.doiorg/10.1787/9789264255340-en.

OECD (Organisation for Economic Co-operation and Development). 2017a. *Agricultural Policies in the Philippines*. Paris: OECD Publishing. http://dx.doiorg/10.1787/9789264269088-en.

OECD (Organisation for Economic Co-operation and Development). 2017b. *Building Food Security and Managing Risk in Southeast Asia*. Paris: OECD Publishing. http://dx.doi.org/10.1787/9789264272392-en.

OECD (Organisation for Economic Co-operation and Development). 2018a. *Innovation, Agricultural Productivity and Sustainability in China*. OECD Food and Agricultural Reviews. Paris: OECD Publishing. https://doi.org/10.1787/9789264085299-en.

OECD (Organisation for Economic Co-operation and Development). 2018b. *Innovation, Agricultural Productivity and Sustainability in Korea*. OECD Food and Agricultural Reviews. Paris: OECD Publishing.

OECD (Organisation for Economic Co-operation and Development) and World Bank. 2014. *Science, Technology and Innovation in Viet Nam*. Paris: OECD Publishing. http://dx.doi.org/10.1787/9789264213500-en.

Official Gazette. 2015. Issue nos 06-08/May 2015. Decision No. 575/QD-TTg.

Pardey, P., C. Chan-Kang, J. Beddow, and S. Dehmer. 2015. "Long-Run and Global R&D Funding Trajectories. The U.S. Farm Bill in a Changing Context." *American Journal of Agricultural Economics* 97 (5): 1312–23.

Pardey, P., J. Alston, and V. W. Ruttan. 2010. "The Economics of Innovation and Technical Change in Agriculture." In *Handbook of the Economics of Innovation*, Volume 2, edited by B. Hall and N. Rosenberg, 939–84. Amsterdam: Elsevier.

Rivera, W. M., and G. Alex. 2004. "Extension Reform for Rural Development." Agriculture and Rural Development Discussion Paper 10, World Bank, Washington, DC.

Spielman, D. J., K. Davis, M. Negash, and G. Ayele. 2011. "Rural Innovation Systems and Networks: Findings from a Study of Ethiopian Smallholders." *Agriculture and Human Values* 28 (2): 195–212.

Stads, Gert-Jan, Haryono, and Siti Nurjayanti. 2007. "Key Trends in Agricultural R&D Investments in Indonesia." International Food Policy Research Institute, Washington, DC.

Stads, Gert-Jan, and Khamphay Manivong. 2006. "Laos." ASTI Country Brief 32, International Food Policy Research Institute, Washington, DC.

Stads, Gert-Jan, and Nguyen Viet Hai. 2006. "Vietnam." ASTI Country Brief 33, International Food Policy Research Institute, Washington, DC.

Stads, Gert-Jan, and Pau Sian Kam. 2007. "Myanmar." ASTI Country Brief 38, International Food Policy Research Institute, Washington, DC.

Stads, Gert-Jan, Sandra Perez, Carlos Covarrubias, and Nienke Beintema. 2016. "Agricultural R&D Indicators Factsheet. Chile." IFPRI, IDB, and INIA, Washington, DC. https://www.asti.cgiar.org/sites/default/files/pdf/factsheets/Chile-Factsheet.pdf.

Suphannachart, W. 2017. "A Review of Agricultural Research System in Thailand: The Development, Policies, Institutions, Investment Patterns, and Impact Assessment." *Review of Integrative Business and Economics Research* 4 (4): 63–89.

Swanson, B., and R. Rajalahti. 2010. "Strengthening Agricultural Extension and Advisory Systems: Procedures for Assessing, Transforming, and Evaluating Extension Systems." Agricultural and Rural Development Discussion Paper 44, World Bank, Washington, DC.

UNCTAD (United Nations Conference on Trade and Development). 2018. *Science, Technology & Innovation Policy Review. Thailand*. Geneva: UNCTAD.

UPOV (International Union for the Protection of New Varieties of Plants). 2020. "Membership." https://www.upov.int/members/en/.

USDA (United States Department of Agriculture). 2018. "Indonesia Agricultural Biotechnology Annual." GAIN Report Number: ID1838, USDA Foreign Agricultural Service.

WEF (World Economic Forum). 2018. *Innovation with a Purpose: The Role of Technology Innovation in Accelerating Food Systems Transformation. System Initiative on Shaping the Future of Food Security and Agriculture*. Geneva: World Economic Forum together with McKinsey & Company.

World Bank, n.d., "Background Report on Vietnam AIS." World Bank, Hanoi.

World Bank. 2005. *Agriculture Investment Sourcebook*. Economic and Sector Work. Washington, DC: World Bank.

World Bank. 2011. "Chile's Agricultural Innovation System: An Action Plan towards 2030." World Bank, Washington, DC.

World Bank. 2012. *Agricultural Innovation Systems. Investment Sourcebook*. Washington, DC: World Bank.

World Bank. 2015. "Sustainable Agriculture Transformation Project." Project Appraisal Document, World Bank, Washington, DC.

World Bank. 2016. *Transforming Vietnamese Agriculture: Gaining More from Less*. Vietnam Development Report. Washington, DC: World Bank.

World Bank. 2017a. *ICT in Agriculture: Connecting Smallholders to Knowledge, Networks, and Institutions*. Washington, DC: World Bank. https://openknowledge.worldbank.org/handle/10986/27526.

World Bank. 2017b. *Myanmar's Public Expenditure Review: Fiscal Space for Economic Growth*. Yangon: World Bank, Myanmar.

World Bank. 2018a. "The Greening of Farm Support Programs: International Experiences with Agricultural Subsidy Reform." World Bank, Washington, DC.

World Bank. 2018b. "The Human Capital Project." In *World Development Report 2019: The Changing Nature of Work*. Washington, DC: World Bank. doi:10.1596/978-1-4648-1328-3.

World Bank. 2019. "*Managing Headwinds. East Asia and Pacific Economic Update*." Washington, DC: World Bank. doi:10.1596/978-1-4648-1412-9.

World Bank. 2020. "The Future of Agriculture & Food Sector. A Scenario Analysis." Unpublished, World Bank, Washington, DC.

Zambon, I., M. Cecchini, G. Egidi, M. S. Saporito, and A. Colantoni. 2019. "Revolution 4.0: Industry vs. Agriculture in a Future Development for SMEs." *Processes* 7 (1): 36. doi:10.3390/pr7010036.

7 Conclusion

A GROWING NEED FOR TRANSFORMATIVE INNOVATIONS

The East Asia region's agriculture and food systems have come under growing pressure to address the economic, environmental, and health-related weaknesses associated with expanding and intensifying production. With economic change, the region's agri-food systems face increasing pressure to meet demand for more food—focusing on nutrition, safety, and convenience. A long-standing emphasis on raising production has had negative effects on the environment, however, contributing to lower agricultural productivity, environmental degradation, zoonoses, and impairment of human health. The COVID-19 (coronavirus) crisis, with its widespread effects on lives and livelihoods, food security, and agri-food system trade, is a hard reminder of the prevailing demand- and supply-side constraints affecting the region's food systems. Several policy and institutional shortcomings, as described in the report and summarized below, limit the sector's ability to invest in and innovate for improved productivity, safety, and sustainability.

Harnessing a new generation of agricultural innovation will be critical to transforming agri-food systems in developing East Asia. Although increased productivity remains an important goal, several transformative innovations, as discussed in the report, hold great potential to deliver a "triple win," that is, to enhance productivity, increase food safety and nutrition, and improve environmental sustainability.

- Innovations supporting agricultural production—for example, biotechnology, urban farming, precision agriculture, climate-smart agriculture, along with e-services—have already demonstrated great potential not only to improve productivity but also to increase agri-food systems' sustainability and resilience to external shocks.
- Innovations along the food chain, including food safety diagnostics, blockchain, the Internet of Things, and e-commerce, have the potential to increase efficiency and food safety and improve health outcomes through improved traceability, reduced food loss and waste, and lower greenhouse gas emissions.

- Innovations related to food consumption and nutrition, including alternative food sources, can improve health outcomes and, in some cases, enhance environmental sustainability and efficiency.

However, the feasibility of adopting transformative innovations is highly context specific, both across and within countries. The ability of farmers and firms to adopt such innovations is affected by the policy incentives that influence the cost of technology, access to finance, land ownership, information and communication technology (ICT) connectivity, and the level of farmers' and firms' knowledge and skills.

Developing East Asian countries exhibit different levels of preparedness for adopting or creating transformative innovations. The ability to adapt, utilize, or create new transformative innovations varies by each country's level of science, technology, and innovation (STI) capacity, as well as by the ease with which the innovations can be adopted in a given agri-food system context. The front-runners in nearly all aspects of transformative innovation are the region's upper-middle-income countries—China, Malaysia, and Thailand—which have relatively high STI and adoption capacity. Several large lower-middle-income countries, including Indonesia, the Philippines, and Vietnam, are well positioned to develop or adopt traditional and new innovations (for example, e-services, climate-smart agriculture, and genetically modified [GM] crops), although their capacity to address traceability, food safety, and sophisticated nutrition innovations is still relatively underdeveloped. Other countries in the region still need to build their STI capacity, but they have the capacity in the short term to increase adoption of some traditional and existing transformative innovations, such as less advanced and less costly e-services, urban agriculture, and precision agriculture.

An agricultural innovation systems (AIS) approach can be instrumental in building countries' STI capacity and farmers' and firms' ability to adopt and implement transformative innovations. An AIS approach can support existing efforts to transform agri-food systems by facilitating more integrated approaches, which can help policy makers identify key constraints to innovation across the agri-food system, as well as the policies and resources needed to alleviate them. It can enable them to combine traditional interventions (support for mostly public research and development [R&D], extension, and education) more effectively with complementary interventions—such as professional skills, co-innovation, private sector innovation, and policy, regulatory, and institutional conditions—needed to support greater innovation.

Promoting innovations in the agri-food sector requires both short- and longer-term measures to strengthen each country's AIS. In the short term, it will be important to reform policies and institutional arrangements to strengthen the enabling environment for agri-food sector innovation and to guide longer-term investments in innovation capacity. Where countries face resource or capacity constraints, short-term measures can begin selectively, focusing on strengthening innovation capacity in specific segments of the agri-food sector or on specific value chains. Selective measures could include, for example, training for firms and farms on produce quality and food safety standards, facilitation of public-private partnerships (PPPs) for applied R&D and e-extension, and targeted infrastructure investments. Longer-term measures to build AIS to enable transformative innovations will require strengthening human capital across the agri-food system. The short- and long-term strategies can be enacted in parallel and should complement each other. For instance, targeted short-term measures focused on specific commodities could be implemented simultaneously with more comprehensive efforts to reform agricultural training and education and demand-driven public-private R&D.

The sections that follow outline complementary spheres of AIS investment in five areas that will be critical to spurring innovation and transforming developing East Asian countries' agri-food systems.

STRENGTHENING INNOVATION POLICY AND GOVERNANCE

Fostering transformative innovations in the region's agri-food sectors will require countries to strengthen their innovation policies. Government STI policies have traditionally focused on productivity and given cursory attention to wider agri-food system demands and challenges. A policy shift toward sustainability and a smaller environmental footprint has taken hold in the region. However, countries' visions and strategies are not fully aligned with investments, particularly with respect to agricultural R&D and innovation. Several countries in the region, including China, Indonesia, Malaysia, Thailand, and Vietnam, have adopted new policy frameworks with a greater orientation toward sustainability, food safety, and the knowledge-based economy. Such strategies can support the transition to more sustainable and resilient agriculture, which responds to growing consumer demand for safer, more nutritious food and facilitates ongoing dietary diversification away from cereals consumption. A few relatively more food-insecure countries in the region still focus largely on productivity increases, albeit with an emphasis on more limited environmental footprints.

Fostering transformative innovations will also require strengthening the governance of AIS. Governance arrangements for AIS vary across developing East Asia. All countries have a dedicated organization (for example, the line ministry for the agriculture sector, or a specialized agency under it) with oversight responsibility for AIS-related matters. However, the level of integration between the AIS and the broader national innovation system (NIS) varies across the region. Evidence reviewed for this report suggests that the institutions that focus primarily on research, and not on innovation more broadly, are often less effective in influencing wider AIS and NIS STI policies than those with an AIS-wide mandate. Coordination across agricultural subsectors (for example, through brokers and platforms) is still limited, apart from commodity boards.

Countries that have moved toward nationally mandated, but independently governed, AIS organizations (such as Japan and the Republic of Korea) are better able to coordinate their agricultural innovation policies, and those policies are more integrated with the broader NIS, which helps ensure that a more coherent set of STI policies, strategies, and activities is defined and funded. Effective processes are generally informed by stakeholder consultations and underpinned by sufficiently resourced program monitoring and evaluation (M&E) efforts. The study suggests that assessment of innovation policies and instruments as part of the governance and decision-making process is still largely missing in all countries reviewed. The review indicates that implementation and institutional ownership of AIS in the region can be strengthened through (1) stronger coordination of agricultural AIS and R&D at the national level, and (2) greater coordination at subnational and subsectoral levels, using

BOX 7.1

Good practices in agricultural innovation policy and agricultural innovation systems governance

The following good practices have been implemented by countries that have more successfully developed their agriculture innovation systems (AIS) policies and governance systems:

- Developing modern, forward-looking agricultural science, technology, and innovation (STI) policies and strategies that extend across the entire agri-food system to enable the transition to more sustainable and resilient agriculture (as in Chile, and in some dimensions, China, Indonesia, Malaysia, Thailand, and Vietnam).
- Deepening existing policies and strategies to encompass food safety, nutrition, zoonoses control, post–farm gate innovations, and food loss and waste to strengthen agri-food systems' ability to cope with emerging risks.
- Establishing a nationally mandated but independently governed inclusive AIS organization (beyond agricultural research and development [R&D]) that coordinates the development of a strategic vision and policies and engages with the broader national innovation system, in which STI policies, strategies, and activities are defined (as in Australia, Chile, Japan, the Republic of Korea, and the Netherlands). Wide stakeholder consultation (for example, through subsectoral AIS platforms) and a consistent focus on monitoring and evaluation should guide the processes.
- Strengthening coordination of agricultural R&D at the national level—particularly in contexts with large and complex R&D and higher education institutional settings—to improve the overall efficiency of R&D and AIS governance (as in Australia and Chile).
- Complementing coordination of AIS at the national level with inclusive subnational and subsectoral coordination (for example, commodity- and theme-specific platforms in Indonesia, Malaysia, and Thailand; industry associations in China) and networking to enhance feedback and stakeholder ownership for implementation (as in Chile, the Netherlands, and Thailand).
- Strengthening and gradually mainstreaming monitoring and evaluation as part of the AIS policy-making and management process (to some extent in Chile), including providing adequate resources for evaluation and enforcing implementation of evaluation practices in relevant government agencies.

processes that include key stakeholders. The countries that have been more successful in developing their AIS policy and governance systems have taken actions along the lines of those highlighted in box 7.1.

REORIENTING AGRICULTURAL R&D TO MEET EMERGING CHALLENGES

Aligning R&D efforts with new and emerging challenges facing the region's agri-food system will be important. Despite significant cross-country differences in R&D allocation, agricultural R&D spending focuses primarily on staple crops (especially rice) and commodities. Moreover, relatively few resources go to the livestock or fruit and vegetable subsectors (just 3–17 percent)—products in ever-greater demand as the region's increasingly affluent consumers diversify their diets. Shifts in consumer preferences, environmental pressures, persistent food safety and nutrition concerns, and the rising risk of zoonotic diseases all indicate that countries' agricultural R&D efforts urgently require adjustment.

Increased public and private financing of agricultural R&D, together with partnerships between public and private entities, will also be key to supporting the region's efforts to respond to emerging needs. Currently, there is still a considerable mismatch between public policy statements and public spending allocations for AIS. At 0.6 percent of agricultural gross domestic product, average spending on agricultural R&D as a share of agricultural output in developing East Asia is well below the level found in Organisation for Economic Co-operation and Development countries (2.5 percent). Based on international experience, public and private agricultural R&D funding of at least 1.0 percent of agricultural gross domestic product is required if countries are to make sustainable progress. There is also a mismatch between public and private sources of R&D finance in agriculture. Although private sector R&D is on the rise, accounting for 25 percent of all agricultural R&D spending in upper-middle-income countries, it is still below the 50 percent share found in Organisation for Economic Co-operation and Development countries.

Private sector R&D remains an important untapped resource for agri-food sector innovation in developing East Asia. Private sector R&D tends to be geared toward modern input technologies, precision agriculture, machinery, e-services, and food-related innovations with commercial potential. Despite recent increases in private sector spending on R&D, institutional and regulatory barriers such as weak enforcement of intellectual property rights (IPR), curbs on private and foreign R&D on plant breeding and GM crop development, and restrictive regulations on agrochemicals, seed, and biosafety, still impede private sector R&D and innovation-related spending. Encouraging increased private sector R&D would thus require addressing such institutional and policy distortions. Several countries in the region have already worked to improve incentives for private sector R&D (for example, tax credits, special economic zones to attract foreign direct investment (FDI), PPPs, and venture capital) or have provided support for capacity building, co-innovation, and commercialization. But further reforms are needed. Fostering the emergence of an innovation entrepreneurial community in agri-food systems will also be critical if countries are to make the transition from technology adopters to generators of cutting-edge innovations.

International collaboration also remains critical to addressing regionwide agri-food system challenges. The region has a long history of collaboration with the International Agricultural Research Centers, and such collaboration remains important. Collaboration networks have centered largely on topics that receive less domestic attention (for example, livestock, farming systems, and the environment) or pose significant cross-country challenges (for example, climate change). Evidence from global R&D reviews indicates that countries that have strengthened and leveraged domestic, regional, and international networks for knowledge exchange and collaboration on such strategic cross-country issues as climate change, natural resource management, and zoonoses are better positioned to respond to emerging needs. In short, countries in the region, even smaller ones with lower innovation capabilities, have multiple channels through which to benefit from agri-food sector R&D: building their domestic R&D capacity, enabling greater private R&D through appropriate policy and regulatory reform, and strengthening long-standing links to international research networks.

The countries that have been more successful in aligning their agricultural R&D system with the AIS have taken the actions set forth in box 7.2.

BOX 7.2

Good practices for agricultural research and development

The following good practices have been implemented by countries that have more successfully aligned their agricultural research and development (R&D) systems with their agricultural innovation systems (AIS):

- Shifting from a predominantly crop and commodity approach toward more diverse and multidisciplinary R&D (as in Chile and the Netherlands).
- Increasing public and private funding for agricultural R&D in a sustained manner, with budgets for agricultural R&D reaching at least 1 percent of agricultural gross domestic product (as in Malaysia). Indeed, R&D investments in the range of 2 percent may be warranted—with a predominant share coming from the private sector—if the countries wish to join the global front-runners in agri-food innovation and sector development (as in Brazil and the Organisation for Economic Co-operation and Development).
- Reallocating funding to agricultural R&D from more-distorting forms of support to the sector (for example, input subsidies); identifying new sources of funding (for example, industry, farmer organizations, commodity levies, and foreign direct investment, as in China, Malaysia, Thailand, and Vietnam on some dimensions); and encouraging cross-sector and cross-country collaboration on strategic innovations to address the persistent underinvestment in agricultural R&D.
- Building domestic R&D and innovation capacity through sustained efforts to strengthen human resources, coordination, and partnerships to enable sustainable sector transformation (as in Brazil and the Netherlands, and in some dimensions, China).
- Providing incentives (for example, tax credits and venture capital in Indonesia, Malaysia, Thailand, and Vietnam); reducing institutional and regulatory barriers for domestic and foreign private sector R&D (for example, foreign direct investment in China); fostering public-private partnerships (as in Chile, and to some extent China and Indonesia); and supporting capacity building, partnership creation, and commercialization (in some dimensions China, Indonesia, Malaysia, the Philippines, and Thailand) to stimulate greater domestic and foreign private sector R&D and innovation.
- Continued strengthening and leveraging of domestic, regional, and international collaboration networks for knowledge exchange and collaboration on strategic cross-country challenges such as climate change, natural resource management, and zoonoses; for smaller countries, strengthening of scientific and knowledge exchanges with regional centers of excellence.

MODERNIZING AGRICULTURAL EXTENSION SERVICES

To increase the impact of agri-food sector innovation on productivity, safety, and sustainability, strengthening and modernizing countries' agricultural extension services will be critical. Extension services in developing East Asian countries are still mostly dominated by the public sector, with variable but increasing diversity of providers, levels of decentralization, and integration of e-extension approaches. Over time, several countries in the region have increased decentralization of their extension services to improve coverage, worked to increase relevance and quality through more demand-driven offerings, and increased the availability and integration of public and private e-extension services. Despite these efforts, concerns remain in many countries about limited coverage and low quality of both public and private extension services, inadequate penetration of e-extension (particularly among lower-middle-income countries), chronic underfunding, and continued reliance on development partners for funding. Most extension systems still tend to be supply-driven, with weak links between knowledge providers and the farms and firms that need it, resulting in levels of technology adoption and innovation among farms and firms that are well below potential.

Agricultural extension services have the potential to increase returns to innovation, but most require reform, better integration of e-extension, and adequate and sustained funding. Evidence from global extension reviews indicates that demand-driven, pluralistic, and decentralized extension systems—ones with adequate coverage and quality, and that integrate e-extension—can better serve the needs of a diverse and often scattered clientele. Efforts to promote such extension systems may require reform of core extension policies, capacity building among both public and private service providers, collective action among smallholders for promoting scale economies, professional management, and systematic demand articulation, as well as financial incentives for private and third-party service providers and for improved ICT connectivity. The countries that have been more successful in upgrading their agricultural extension systems have taken the actions outlined in box 7.3.

BOX 7.3

Good practices for policies on agricultural extension services

The following good practices have been implemented by countries that have more successfully upgraded their agricultural extension services:

- Ensuring adequate and sustained funding to develop more demand-driven, pluralistic, and decentralized extension systems (that is, improved coverage, content, and quality, realized in some dimensions in China, Indonesia, the Philippines, and Vietnam) with integration of e-extension to improve the returns to agricultural innovation (for example, e-centers in China).
- Adjusting policies to enhance pluralism and enable greater private and third-party service delivery (reforming laws, strengthening financial incentives, increasing capacity).
- Increasing integration of e-extension, for example, through improved information and communication technology connectivity and farmer capacity, price incentives for information and communication technology suppliers, and dedicated training programs to upgrade the skills of public and private extension agents and farmers (for example, e-training in China, the Philippines, and Thailand).
- Making extension services more demand-driven and innovation processes more inclusive by facilitating collective action among smallholders (as in China, Indonesia, and Thailand); setting up platforms and brokers (as in Chile, and to some extent China, Malaysia, and Thailand); establishing technology incubators and centers for technology transfer at the local and provincial levels (as in Vietnam) and other organizational forms of extension-research-farm or -firm links that systematically articulate farmer and firm demand for and feedback on knowledge and innovations (for example, technology demonstrations, cooperatives, and industry associations, as in China and Indonesia).

BUILDING CAPACITY AND SKILLS FOR AGRI-FOOD SYSTEM INNOVATION

A lack of adequate skills among those working in the agri-food system is one of the key bottlenecks to adoption of innovations and the development of the AIS in many developing East Asian countries. Although the nature and quality of tertiary and technical and vocational education and training (TTVET) systems vary significantly across the region, they have several common characteristics. Agriculture TTVET systems suffer from declining numbers of graduates because they face difficulties in attracting both students and qualified lecturers. In addition, countries' vocational training systems are often unable to respond to the labor market demand for the skills (for example, business management, ICT) needed in modern agri-food systems. Moreover, cross-disciplinary skills and collaboration are limited, which constrains the capacity for co-innovation on several important cross-sectoral issues, including zoonoses, nutrition, and food safety. The deficiencies in the training of workers by firms and in the education and skills of aging farmers limit necessary knowledge upgrading and undermine progress toward higher-value production, productivity, and environmental sustainability. In response, several promising reforms have been carried out in upper-middle-income and high-income countries in the region to address the relevance and attractiveness of TTVET, agri-food sector skills shortages, and the lack of cross-sectoral collaboration.

Innovation capacity and skills are an essential precondition for long-term sustainability of agri-food systems in developing East Asia. The findings of the review indicate that continued efforts to create a skilled workforce for future agri-food systems will require modernization of the TTVET systems in many developing East Asian countries. The region also needs skilled workers capable of collaboration across disciplines and between industry and public agencies to better respond to emerging challenges. The region's workforce in agribusiness and on farms is in dire need of improved knowledge and skills, tailored to countries' specific circumstances and needs. For example, although farmer capacity can be improved through agricultural extension and training programs, relevant, demand-driven TTVET training becomes increasingly important when countries undergo industry-led agri-food system transformation. The slow pace at which skills are built highlights the need for immediate action. The countries that have been more successful in developing their TTVET systems and upgrading their workforce skills have taken the actions summarized in box 7.4.

BOX 7.4

Good practices for agricultural education and training

The following actions have been taken by countries that have more successfully developed their tertiary and technical and vocational education and training and upgraded their workforce skills:

- Building a pool of skilled workers across the agri-food system through longer-term reforms of the tertiary (China) and vocational education systems (the Republic of Korea and Singapore).
- Developing more demand-driven curricula (content, pedagogy, cross-disciplinary courses, inclusion of industry views, industry internships) that better match the needs of rapidly changing

continued

Box 7.4, *continued*

agri-food systems (as in Australia and Singapore, and on some aspects China, Malaysia, and Thailand).

- Establishing a permanent multistakeholder consultative structure that periodically reviews the relevance of the curricula to sector and industry needs (as in Australia and the Netherlands).
- Enhancing collaboration between tertiary and technical and vocational education and training institutions, industry, and public agencies to improve agri-food systems' ability to respond to emerging risks and opportunities, including through university-industry linkage programs (as in Australia, China, Malaysia, and Thailand) or centers of excellence (as in China, Malaysia, the Netherlands, and Thailand) that facilitate joint problem solving (for example, in new or underfunded areas, such as food safety, livestock production, and nutrition) and financing.
- Enhancing the standing of vocational training (as in Malaysia and Thailand).
- Providing opportunities for upgrading the skills of agri-industry employees, for example, through vocational training (as in Singapore).
- Strengthening farmer education and skills to improve productivity in a safe and sustainable manner. Extension and farmer training programs may best suit smallholder contexts (as in China and Thailand). Technical and vocational training for farmers becomes increasingly important in countries undergoing industry-led agri-food system transformation (as in Australia, Korea, and the Netherlands).

STRENGTHENING THE ENABLING ENVIRONMENT FOR AGRICULTURAL INNOVATION

The enabling environment for agricultural innovation needs continued strengthening across developing East Asia. An enabling environment—especially with respect to IPR protections, food and biosafety regulations, and infrastructure development—is critical to supporting innovation in agriculture. Adequate IPR protection is particularly important with respect to GM crops, enabling private R&D and investment, including via FDI. Although most countries in the region have IPR regulations and laws in place, differences exist in the implementation of such regulations, for example, on plant variety protections. Moreover, inadequate IPR protection has been found to impede R&D investment (public, private, and foreign) in GM crops. In addition to IPR, well-defined and consistently applied biosafety regulations can provide a powerful stimulus for investment in biotechnology. To date, most countries in developing East Asia have developed, or are in the process of developing, regulatory systems to manage biosafety and enable innovation in GM crops. Enforcement of biosafety regulations still faces difficulties, however, and there is limited application of GM crops in several countries.

However, emerging issues facing the region's agri-food systems are creating new demand for an adequate enabling environment. Food safety has emerged as a critical topic that is high on the agenda of consumers, governments, and the agri-food industry alike. All developing East Asian countries have food safety regulations in place; however, the countries' capacity to respond to new issues and to enforce the regulations varies significantly. The more urbanized upper-middle-income countries have modernized their food safety systems, but most other countries covered in this study have difficulty meeting the food safety requirements of their fast-changing food systems. Continuing gaps between food safety capacity and requirements are especially challenging in rapidly urbanizing middle-income settings.

Efficient and well-developed infrastructure, access to finance, and land rights play a critical role in connecting firms and farms to market opportunities and enhancing their ability to invest in and adopt innovations. Although the upper-middle-income countries have mostly good-to-excellent logistics and ICT infrastructure, lack of adequate productive infrastructure in several lower-middle-income countries still limits farm and firm potential. In most countries, financial services and instruments available to farmers and firms are an investment-limiting factor. Moreover, inadequate incentives for innovation-friendly FDI still prevent most countries from benefiting from foreign investment and innovation in agriculture. Reliable land rights and functioning land markets can improve access to finance and enable innovation. Indeed, many countries in the region have carried out consecutive land reforms that enable land to be used as collateral and facilitate land consolidation, rental, and sale. However, complex registration, allocation, and ownership of land rights still impede access to finance and investment by farms and firms in several countries. The countries that have been more successful in developing their enabling environment for agricultural innovation have taken the actions outlined in box 7.5.

BOX 7.5

Good practices for strengthening the enabling environment for innovation

The following actions have been taken by countries that have more successfully developed their enabling environment for agricultural innovation:

- Strengthening incentives that reward innovation by continuing to align intellectual property rights (IPR) protection laws with international standards and by continuing to improve enforcement of IPR (as in the Organisation for Economic Co-operation and Development).
- Establishing and improving biosafety frameworks and strengthening enforcement of existing biosafety regulations to maximize the potential of biotechnology to improve productivity and the resilience of the sector (as in Brazil).
- Strengthening food safety regulations and capacity, beginning with the most critical risks in the agri-food system, including via risk-based surveillance systems, is a condition for adoption and invention of many food safety innovations (as in the Republic of Korea and Singapore).
- Improving the quality of infrastructure and information and communication technology connectivity in rural areas to enable private sector market opportunities, investment, and innovation, and enhance the potential of e-commerce, e-services, precision agriculture, and other digitally enabled activities (to some extent, China).
- Broadening the range of financing options available to farmers and firms (to some extent, Malaysia and Thailand); gearing instruments for innovation finance (for example, research and development tax credits, competitive innovation funds, venture capital) and other support for innovation-related investments (for example, incubators, support to matchmaking, and IPR management) toward greater private sector engagement.
- Improving regulatory and institutional frameworks that govern agricultural land market operations in the rural economy to enhance long-term investments in agricultural productivity and sustainable growth (as in China and Thailand).

The agri-food system in developing East Asia has reached a critical point. The system has provided food and employment to growing populations through innovations and economic policies that have been conducive to agricultural productivity growth, but that have come with significant cost. The region can no longer afford the costs to human capital and the economy of environmental degradation, unsafe food, persistent nutrition deficiencies, and zoonotic diseases. Further investments in agricultural STI and in farmers' and firms' capacity to adopt new knowledge and emerging technologies—centered on the areas discussed above—are urgently needed in a region that relies on millions of smallholders for their agri-food outputs and services.

There is no one-size-fits-all policy mix. Indeed, given the substantial heterogeneity across countries, policy actions will need to be tailored to specific country circumstances. Table 7.1 provides an illustrative summary of the types of innovations and related policies that could enable greater agricultural innovation in countries covered in this report, while keeping in mind that deeper country-level analysis would be needed for development of appropriate country-specific strategies and policy recommendations. One thing is certain: Fostering greater innovation in agriculture will require concerted action. Countries that build strong AIS capacity, exhibit strategic vision and leadership, provide a favorable policy environment, and provide adequate support for R&D, extension, and skills upgrading can foster the necessary transition to more resilient and sustainable agri-food systems.

TABLE 7.1 Summary of the transformative innovations with high potential for country readiness and the agricultural innovation systems policy areas warranting attention in China, Indonesia, Malaysia, the Philippines, Thailand, and Vietnam

COUNTRY OR SECTOR	TRANSFORMATIVE INNOVATIONS WITH POTENTIAL[a]	COORDINATION AND GOVERNANCE	AGRICULTURAL RESEARCH, EXTENSION, AND INCENTIVES FOR PRIVATE SECTOR AND PARTNERSHIPS	INNOVATION SKILLS	ENABLING ENVIRONMENT
China	**Biotechnology, precision farming**, urban agriculture, **e-services,** food safety, traceability, **sustainable agriculture practices**, food loss and waste, **e-commerce**	• Strengthen inclusive coordination for AIS • Improve strategic AIS leadership and mechanisms for province to state level coordination • Consistent use of M&E for policy and priority setting	• Increase spending and cross-cutting R&D on emerging issues • Limit public R&D on developmental research to enable private sector investment • Further enable PPP, also by foreign firms • Strengthen decentralization and coverage of extension; greater focus on sustainable practices and business support • Enable brokers to support innovation processes	• Strengthen capacity for cross-disciplinary collaboration • Improve ICT skills in rural areas	• Enforce IPR, biosafety, and environmental governance • Improve ICT connectivity • Create an ecosystem for venture capital; greater flexibility with FDI rules
Indonesia	Varietal development, CSA, **biotechnology,** precision farming, e-services, food safety and traceability	• Strengthen use of M&E for policy and priority setting • Support visionaries and platforms for inclusive innovation	• Increase R&D spending and reduce commodity focus to allow multidisciplinary R&D • Increase incentives for co-innovation • Improve quality of extension at the decentralized level, including with e-extension	• Improve ICT skills in rural areas • Enhance industry-higher education collaboration to address shortage of skills • Implement vocational training reform	• Consider endorsing UPOV • Enforce IPR • Improve transport and ICT infrastructure • Address land tenure and rural finance • Increase incentives for FDI
Malaysia	**Precision and urban farming**, biotechnology, nutrition, food safety and traceability, food loss and waste	• Consistently use M&E for policy and priority setting • Support visionaries and platforms or brokers for inclusive innovation in agriculture	• Maintain or increase spending and targeted allocation (less commodity-focused) • Further enable co-innovation among public and private actors • Improve extension coverage	• Strengthen capacity for cross-disciplinary collaboration • Enhance industry-research-higher education collaboration for greater capacity and to address shortage of skills	• Consider endorsing UPOV • Reform land tenure that impedes investment by farms and firms
Philippines	**CSA**, e-services, **biotechnology**, food safety, precision farming	• Improve coordination and fragmented governance of AIS • Integrate M&E • Support visionaries and platforms for inclusive innovation	• Continue increasing R&D spending and efficiency • Provide incentives for co-innovation through innovation funds and PPPs • Increase budget and content (sustainable practices) of extension and e-extension services	• Improve ICT skills in rural areas • Enhance industry-higher education collaboration to address shortage of skills • Implement vocational training reform	• Consider endorsing UPOV • Improve IPR enforcement • Improve transport and ICT infrastructure • Improve land tenure and rural finance options

continued

TABLE 7.1, *continued*

COUNTRY OR SECTOR	TRANSFORMATIVE INNOVATIONS WITH POTENTIAL[a]	COORDINATION AND GOVERNANCE	AGRICULTURAL RESEARCH, EXTENSION, AND INCENTIVES FOR PRIVATE SECTOR AND PARTNERSHIPS	INNOVATION SKILLS	ENABLING ENVIRONMENT
Thailand	CSA, nutrition, **precision farming,** biotechnology, **e-services,** food safety and traceability	• Strengthen fragmented R&D coordination • Support visionaries and platforms or brokers for inclusive innovation	• Increase R&D spending, reduce emphasis on commodities • Increase incentives for co-innovation • Develop a demand-driven extension model • Improve relevance of extension advice	• Strengthen ICT skills in rural areas and extension services • Continue strengthening capacity for cross-disciplinary collaboration and industry-higher education collaboration to address shortage of skills	• Consider endorsing UPOV • Establish biosafety regulations • Reform land tenure • Improve ICT connectivity
Vietnam	**CSA,** livestock research, **precision farming**, **biotechnology**, food safety and traceability	• Strengthen coordination of R&D and AIS strategies and programs • Strengthen use of M&E • Support visionaries and platforms for inclusive innovation	• Increase R&D spending, further strengthen coordination across agencies and actors • Enable increase in private sector R&D and provide incentives for co-innovation (funds, PPPs) • Improve coverage, budget, and plurality of extension services	• Improve ICT skills in rural areas • Strengthen the HR pool for R&D and other skills • Invest in HR for agriculture 4.0 • Enhance industry-higher education collaboration to address shortage of skills • Implement vocational training reform	• Enforce IPR, biosafety, and environmental governance • Improve transport and ICT infrastructure • Improve financing options for firms and farms
Role of public and private sector	Context specific. See chapter 5 and table 5.1.	• Public leadership is necessary at the national level, strategic issues, and M&E. Private sector leadership more prominent for, for example, specific value chains. Inclusive participation is the norm.	• Context specific. See chapter 6 and box 6.7.	• Public sector leadership and investment is the norm. However, private sector inputs to curriculum development, industry-university collaboration, and internship is crucial.	• Public sector leadership and investment is the norm in most contexts. Private sector interest in infrastructure and development of regulations should be encouraged.

Source: Original table for this publication.
Note: The table is illustrative only, and expected to provide a basis for further work and investigation to help refine countries' strategic attention to transformative innovations and AIS. AIS = agriculture innovation systems; CSA = climate-smart agriculture; FDI = foreign direct investment; HR = human resources; ICT = information and communication technology; IPR = intellectual property rights; M&E = monitoring and evaluation; PPP = public-private partnership; R&D = research and development; UPOV = International Convention for the Protection of New Varieties of Plants.
a. Transformative innovations with potential (chapters 4–5). The innovations in bold are already priorities according to the literature review. AIS elements are drawn from chapter 6 and from tables F.1 and F.3 in appendix F.

APPENDIX A

Plastic Use and Management in East Asia

AGRICULTURAL PLASTIC USE

Plastic use and waste, along with inadequate recycling, have generated marine debris and become a major environmental problem constraining the green development of agriculture. Plastic is used widely in agriculture: films in greenhouses, walk-in tunnel and low tunnel covers, mulching and silage; nets for protection from birds, insects, and hail; strapping for bales; irrigation pipes; and fertilizer and agrochemical packages (Schuttelaar and Partners 2019; World Bank 2019). At present, bottlenecks on plastic reconversion and reuse are found in the high variety of mingled plastics and the cost associated with identifying and separating the different streams. Leading companies acknowledge the need to develop innovative business models for addressing the issue of marine litter. However, innovations in both technologies and financing happen in an ad hoc manner, often not to scale or not adaptable to local contexts, and with lack of opportunities for comparisons, quality control, and feasibility (World Bank 2019). In addition, the food and beverages sector relies increasingly on plastic packaging. Table A.1 summarizes plastic use, mismanagement, recycling, and main plastic sources for East Asian countries.

TABLE A.1 **Plastic generation, mismanagement, recycling rate, and main sources in developing East Asia**

COUNTRY	GENERATION OF PLASTIC WASTE (MILLION TONS PER YEAR)	MISMANAGED PLASTIC (MILLION TONS PER YEAR)	RECYCLING RATE (%)	PERCENT OF TOTAL PLASTIC WASTE THAT IS MISMANAGED	MAIN PLASTIC SOURCES
Cambodia	0.4	—	19.5	—	Plastic bags, tourism
China	59.0–61.0	17.2	22.0	27.7	Packaging, including food and beverages, increasingly agriculture
Indonesia	—	3.2	10.0–15.0	10.1	Packaging, 70 percent for food and beverages
Lao PDR	0.01–0.03	—	8.7	—	Tourism, bottles
Malaysia	1.7	—	0.1	2.9	Packaging
Philippines	2.7	1.8	28.0	5.9	Small packages by urban households
Thailand	4.7	1.8	20.8	3.2	Packaging
Vietnam	—	1.8	10.0	5.8	Packaging, especially food and beverages

Sources: Original table for this publication. Data on Malaysia, Thailand, Indonesia, Philippines, Vietnam, Lao PDR, and Cambodia from World Bank 2020. Data on China from Halder 2019; Kwong 2019; Ritchie and Roser 2018. Data on percent of mismanaged plastic waste from Jambeck et al. (2015).
Note: — = not available. Recycling rate for Cambodia is for Phnom Penh only. Mismanaged waste is the sum of inadequately managed waste plus 2 percent littering. Total mismanaged plastic waste is calculated for populations within 50 kilometers of the coast.

REFERENCES

Halder, Ben. 2019. "A Plastic Revolution Is Brewing. . . in China." OZY, April 21, 2019. https://www.ozy.com/around-the-world/a-plastic-recycling-revolution-is-brewing-in-china/93680/

Jambeck, J. R., R. Geyer, C. Wilcox, T. R. Siegler, M. Perryman, A. Andrady, R. Narayan, and K. L. Law. 2015. "Plastic Waste Inputs from Land into the Ocean." *Science* 347 (6223): 768–71.

Kwong, Luke. 2019. "10 Facts about Plastic Waste in Southeast Asia." The Borgen Project, July 12, 2019. https://borgenproject.org/10-facts-about-plastic-waste-in-southeast-asia/.

Ritchie, Hannah, and Max Roser. 2018. "Plastic Pollution." Our World in Data, University of Oxford. https://ourworldindata.org/plastic-pollution.

Schuttelaar and Partners. 2019. "Report on Use of Plastics in Agriculture." Wageningen, the Netherlands.

World Bank. 2019. "Marine Plastic Debris. A Regional Engagement Framework for Action for the World Bank East Asia and Pacific Region." Unpublished, World Bank, Washington, DC.

World Bank. 2020. "Marine Plastics." Separate briefs for Cambodia, Indonesia, Lao PDR, Malaysia, Philippines, Vietnam, Thailand. Unpublished, World Bank, Washington, DC.

APPENDIX B

Management of Zoonoses and Other Emerging Infectious Diseases: Lessons Learned

Livestock production in East Asia has expanded significantly through increasing land area and diversity of livestock production systems and is the key driver for increased incidence of emerging infectious diseases (EIDs). Today, the evolutionary pressures on pathogens present an expanding array of risks to human health and the economy (Allen et al. 2017; Jones et al. 2008; World Bank 2010). A range of human activity is driving EIDs: the hunting and selling of wildlife in markets; the keeping of livestock in unsanitary conditions close to food preparation; co-location of multiple species in unnatural, captive conditions for sale; broader trading in wildlife or endangered species (even when farmed); indiscriminate land-use change in high-risk regions; and indiscriminate use of antibiotics in agriculture (McAleenan and Nicolle 2020). Farming near rainforests, the consumption of bushmeat, and ecotourism are examples of the types of interactions that can create opportunities for pathogens to "jump" species (World Bank 2010).

A holistic interdisciplinary approach to management of EIDs is required. Recognition of the interrelatedness of the respective health domains and of the risks that zoonotic diseases represent to public health has led to appeals for more horizontal interaction among the disciplines and the sector agencies, departments, and ministries that are responsible for public health, medical professions, veterinary services, and the environment. Animal health and food safety are the key pillars of One Health (which assumed urgent practical significance in late 2003 with the emergence of highly pathogenic Asian avian influenza A [H5N1] virus), which is an approach to designing and implementing programs, policies, legislation, and research in which multiple sectors communicate and work together to achieve better public health outcomes. It is particularly relevant for issues such as food safety, the control of zoonoses, and combating antimicrobial resistance. The World Health Organization works closely with the Food and Agriculture Organization of the United Nations and the World Organization for Animal Health to promote multisectoral responses to food safety hazards, risks from zoonoses, and other public health threats at the human-animal-ecosystem interface (WHO, n.d.). Although the general concepts are well accepted, how to implement the One Health concept is still not clearly understood (World Bank 2010). The lessons learned from One Health and several other national and international approaches are summarized in table B.1. They are organized around the pillars of prevention, detection, response, and recovery (Carlin et al. 2019).

TABLE B.1 **Convergence of One Health with several national and international approaches to managing emerging infectious diseases and other biothreats**

ONE HEALTH APPROACH: LESSONS LEARNED AND FOCUS OF ACTION	NATIONAL AND INTERNATIONAL ZOONOSES MANAGEMENT APPROACHES: GAPS IDENTIFIED AND LESSONS LEARNED
Overall findings	
One Health approaches the issue of controlling infectious diseases from a "human-animal-ecosystems perspective." One Health is a collaborative, multidisciplinary, and multisectoral approach that can address urgent, ongoing, or potential health threats at the human-animal-environment interface at subnational, national, global, and regional levels. This approach includes ensuring balance and equity among all the relevant sectors and disciplines.	**Poor coordination among global initiatives.** No strategic interinstitutional guiding framework attempts to align all global initiatives toward a commonly defined objective. **Cross-cutting functions provide underutilized entry points for participation.** Critically underemphasized were community engagement; risk communication and education; research and development in areas beyond surveillance or medical countermeasures; and data and information management.
Awareness, prevention, and protection: Systems, policies, and procedures to determine, assess, avoid, mitigate, and reduce threats and risks by reducing vulnerability and exposure	
Public awareness of the risks of zoonotic diseases and political commitment to containing them tend to fade over time, and other priorities such as financial and food price crises and climate change become more prominent. The continuity of resources devoted to disease surveillance and control has therefore remained an enduring concern. **Strengthening education.** A review of curricula, with more emphasis on epidemiology and the wider effects of ecosystems on human and animal health, is needed to shift human and veterinary health services from controlling to preventing diseases. **Activities driving EIDs:** Intensified farming and concentration of animals, pressure on food production systems, and increasing global movement of people, animals, and animal products have led to evolutionary pressures on pathogens that present an expanding array of risks. Destabilization of ecosystems: (1) destruction and fragmentation destroy the balance between different species and can enable individual species to become dominant; (2) increased interaction between human and wild ecosystems, for example, through farming near rainforests, the consumption of bushmeat, and ecotourism, gives rise to more opportunities for the exchange of pathogens, including transmission to "naïve" or unprotected individuals.	**Lack of sanctions.** None of the management approaches have direct sanctions that can be applied via trade or other means to enforce adherence to international agreements, for example, on reporting requirements of WTO and WHO members. There is a strong rationale to place such risks within trade and other frameworks to "price in" their costs. **Emphasis on managing outbreaks rather than preventing them.** A range of human activity raises the risks of zoonotic disease spillovers. Hunting and selling of wildlife in markets; livestock in unsanitary conditions close to food preparation; co-location of multiple species in unnatural, captive conditions for sale; broader trading in wildlife or endangered species (even when farmed); indiscriminate land-use change in high-risk regions; indiscriminate use of antibiotics in agriculture. Despite this understanding, prevention is scarcely addressed; international systems wait for outbreaks to occur before treating them as health hazards.
A permanent body that coordinates the preparation and regular update of contingency plans to deal with the eventuality of an outbreak. Coordinating authority conferred as a function of executive office, such as a prime or deputy minister, who is served in this capacity by an advisory committee that operates with his or her authority. **Special One Health teams,** composed of representatives of the human, animal, and ecosystem institutions, with responsibility for diseases at the animal-human-ecosystem interface.	**Lack of a single treaty, institution, or forum for focusing world resources on zoonotic disease risks.** Zoonotic disease covers several areas, especially health, ecology, veterinary sciences, and agriculture. As such, it tends to fall between the respective agencies covering these issues. **Dispersed mandate**. At both a national and international level, there tends to be a disparate spread of organizations with some reference to the issue, but no single legislative or institutional focus.
Creation of an independent agency for public health, including zoonoses and food safety. The prospective institutional architecture for global surveillance and control is also considered.	**Intersectoral committees and networks usually dissolve or are hard to maintain.** As soon as the risk or the funding vanishes, government institutions and scientific disciplines often divide up the responsibility and expertise in silos and multidisciplinary collaboration remains weak. Many countries find it hard to maintain joint governance structures, sustain joint funding, or measure its impact.

continued

TABLE B.1, *continued*

ONE HEALTH APPROACH: LESSONS LEARNED AND FOCUS OF ACTION	NATIONAL AND INTERNATIONAL ZOONOSES MANAGEMENT APPROACHES: GAPS IDENTIFIED AND LESSONS LEARNED
Surveillance and detection: Systems, policies, and procedures to gather and analyze information, provide early warning, and inform strategies	
Importance of early detection. One of the most essential factors in the control of any new emerging health risk is early detection of the disease and understanding of its epidemiology. **Delays with detection and reporting.** Limited human capacity and physical facilities and the piecemeal nature of work by public health, veterinary, and environmental agencies acting in isolation from one another along narrowly sectoral lines often cause new EIDs to remain unnoticed. Reporting is often delayed out of fear of the economic losses, for example, from trade bans and reduced tourism.	**Underreporting, absence of reporting, and misreporting of zoonotic outbreaks.** Under the WTO's Sanitary and Phytosanitary Agreement, members must report zoonotic outbreaks that may affect trade to the OIE. However, 62 percent of countries had not submitted a report to the OIE, and many had misreported. Reporting requirements and standards need improvement. **Overreliance on untrustworthy, impermanent, and ineffective national regimes and inability of global mechanisms to monitor and inspect.**
Consultation on priority setting. Developing capacity for risk analysis, often an area of weakness, is a necessary condition for more effective priority setting that aims to identify "hot spots," that is, contexts in which climatic, social, and economic conditions, including the state of sanitation infrastructure and services and the proximity of humans and animals, provide a particularly favorable environment for diseases to emerge or reemerge.	**Lack of a central database for mapping risk areas.** There is a need to create a map of areas where ecological factors, agricultural systems and food supply chains, human economics, climatic conditions and trends, cultural factors, health systems, and past outbreaks combine to create higher risks. While individual institutions and researchers have attempted this, a more comprehensive methodology on which policy decisions can be made requires complex modeling and concerted scientific and political efforts, including access to open data.
Multidisciplinary collaboration, staff exchange, and a conducive working environment. Effective prevention and control measures against emerging and reemerging infectious diseases will require multisector strategies and active collaboration across professional disciplines. Certain select public health activities could be shared, for example, in surveillance by human and animal health field agents. **Joint preparedness planning.** Preparation of integrated national action plans has shown the potential gains from joint planning exercises between public health and veterinary services. More attention is still needed to reduce transaction costs and to ensure that these plans are and remain realistic and implementable.	**Most systems are designed and managed by human health specialists,** not ecologists or veterinarians. Some approaches, such as the WHO's One Health, are correct but not effective enough. Much greater focus is needed on addressing the pre-spillover conditions increasing risks. **Collaboration across disciplines and with local communities is a must to gain impact.** There are fewer barriers to work across disciplines at the local level, which makes it easier to engage with communities and include them in the surveillance and control process. **Need for resources, participatory process and learning by doing, and mentoring to support multistakeholder governance, surveillance, and control.** Experience from Vietnam (antimicrobial resistance governance) points out the need for full adhesion of stakeholders and the provision of appropriate resources to improve the national surveillance system and break existing silos.
Response: Systems, policies, and procedures aimed at controlling or mitigating the impact of disease and saving lives	
Communicating consistent messages. Human and veterinary health channels to communicate information about disease outbreaks need to be harmonized and coordinated. The public, and in particular the different participants in the animal source food chain, need to clearly understand the rationales behind existing disease control strategies and the level of emergency status an EID is assigned at any given time. They also need to understand the safety issues involved in animal products that originate in affected areas. **Funding requirements.** Funding through conventional time-bound, project-based investments is inadequate for response; more reliable, sustained flows of financial resources will need to be established for prevention, detection, and response. Combining surveillance and control systems that focus on pandemic and other common diseases is a feasible option.	**Planning and implementation are dominated by the human health sector.** A tendency to think about biothreats in terms of human health drives planning and implementation processes, even though dozens of sectors are relevant for prevention, detection, response, and recovery. **Disease prevention and control strategies require full understanding of decision making among the actors at the commune and value chain levels, and inclusion of social scientists as part of the analysis.** In the face of large epidemics, the way decisions are made by farmers, traders, transporters, and all stakeholders along the value chain must be analyzed and brought into the equation.
Recovery: Systems, policies, and procedures to restore and strengthen normal operations	
Not covered.	**Recovery is all but missing.** Recovery functions remain the most significantly overlooked. Initiatives to meaningfully and systematically advance recovery planning and implementation are almost nonexistent. Recovery is supported by only 5 of 22 initiatives.

Sources: Original table for this publication. Carlin et al. 2019; Martin 2019; McAleenan and Nicolle 2020; World Bank 2010.
Note: EID = emerging infectious disease; OIE = World Organization for Animal Health; WHO = World Health Organization; WTO = World Trade Organization.

The approaches converge on a number of issues: human activity driving EIDs; the importance of early detection and of clear communication, coordination, and governance by independent dedicated organizations; and the need for multidisciplinary teams for prevention, joint preparedness and priority setting, and implementation of interventions.[1] Reviews of different approaches also highlight a number of gaps: although prevention is discussed, it is rarely addressed; local communities and full understanding of motivations and decision making by key local stakeholders are not necessarily integrated with strategies and interventions; and coordination organizations are challenging to maintain, contributing to less-than-ideal interventions for prevention, detection, and response. Apart from the lack of prevention of EIDs, recovery interventions are also missing (for example, recovery is not a part of One Health), suggesting that the lack of recovery interventions may contribute to persistent human risky activity in "hot spots," and the re-emergence of existing EIDs.

NOTE

1. Carlin et al. (2019) also identify the following cross-cutting functions in building resilience to biothreats—governance (leadership, policy, statute, regulation, enforcement), resource allocation and coordination, community engagement and resilience, risk communication and education, workforce development and sustainment, research and development, and data and information management.

REFERENCES

Allen, T., K. A. Murray, C. Zambrana-Torrelio, S. S. Morse, C. Rondinini, M. Di Marco, N. Breit, K. J. Olivial, and P. Daszak. 2017. "Global Hotspots and Correlates of Emerging Zoonotic Diseases." *Nature Communications* 8: 1124. https://www.nature.com/articles/s41467-017-00923-8.

Carlin, E., C. Machalaba, F. Berthe, K. Long, and W. Karesh. 2019. *Building Resilience to Biothreats*. New York: EcoHealth Alliance. 10.13140/RG.2.2.26016.64001.

Jones, K., N. G. Patel, M. A. Levy, A. Storeygard, D. Balk, J. L. Gittelman, and P. Daszak. 2008. "Global Trends in Emerging Infectious Diseases." *Nature* 451: 990–93. https://www.nature.com/articles/nature06536.

Martin, Vincent. 2019. "Disease, Politics and Food Security: Rethinking Animal Health." https://www.linkedin.com/pulse/disease-politics-food-security-rethinking-animal-health-martin/.

McAleenan, B., and W. Nicolle. 2020. "Outbreaks and Spillovers. How the UK and International Community Can Lower the Risks of Zoonotic Disease." Policy Exchange, London.

WHO (World Health Organization). n.d. "Managing Public Health Risks at the Human-Animal-Environment Interface." World Health Organization, Geneva. https://www.who.int/zoonoses/en/.

World Bank. 2010. *People, Pathogens, and Our Planet. Volume 1: Towards a One Health Approach for Controlling Zoonotic Diseases*. Washington, DC: World Bank.

APPENDIX C

Agricultural Innovation Systems Framework

Figure 3.1 highlights that farms and the firms that may be value chain leaders are the critical players whose decisions on accumulating capital, labor, or knowledge need to be jointly considered. The figure distinguishes the demand for factors from the supply to highlight that, without farm and firm demand, supply-side policies to generate or disseminate relevant knowledge are equivalent to pushing on a string. The division between the two sets of variables is not so sharp, particularly in the knowledge area, and the bi-directional arrows crudely capture the feedback relationship between farms or firms and knowledge institutions.

DEMAND

Incentives to accumulate. The demand side comprises the overall set of incentives to invest and accumulate, including the macro context, in particular, the volatility of sales, the competitive structure, and the trade and investment regime that determine whether leader firms will seek to enter and grow, and farms invest in new technologies. Issues of market access play a particularly important role, both domestically and externally. In addition, demand-related initiatives such as the development of or connection to digital platforms that reduce searching, matching, and informational transaction costs; the establishment of domestic or international commercial network; or even procurement policies are included here.

Education and capabilities. The second set of variables captures farm capabilities: basic educational skills and managerial competencies and training brought by extension that enable a farmer to recognize an opportunity and act to take advantage of it. There are clear interactions between the sets of variables. The ability to participate in a large international market increases the likely benefits of upgrading and innovating, while better capabilities permit farmers to take advantage of markets.

SUPPLY

On the supply side, sources of *human capital* include the entire set of institutions ranging from primary school to technical institutes to universities that provide basic skills to farmers, or advanced training to extension workers and scientists. On the *physical capital* side are efficient domestic industries, easy access to imported capital and intermediate goods, and the contributions of value chain leaders.

The research and extension system. This set of knowledge encompasses the institutions that support farms, including the extension and other services that disseminate new technologies or best practices. The science, technology, and quality systems specifically facilitate technological transfer, adapt existing knowledge, or generate new knowledge. Investments in national food safety systems can play an important role in closing the gap between domestic-traditional and international-modern production standards. Well-functioning national food safety systems reduce compliance costs and attract agri-food value chain investments (Fuglie et al. 2020; Townsend et al. 2018). Actions to improve the food safety system encompass changing regulations, building organizational capacity for inspection and enforcement, setting up laboratories, and investing in education and training. The international innovation system generates most new knowledge. It becomes critical to technological transfer because, particularly in cases in which crops and conditions are unfavorable to private sector provision of research, many of these institutions are nonmarket (government research institutes, universities, and so on). However, private providers of knowledge are increasingly important, and the system needs to encourage their entry, including through intellectual property rights, an efficient trade regime that permits easy import and export, and a generally supportive business environment.

BARRIERS TO ACCUMULATION

The center panel of figure 3.1 captures barriers to the allocation and accumulation of all factors: barriers to land and labor reallocation, missing finance and risk mitigation markets, entry and exit barriers, and poor regulatory measures. Clearly, innovation-specific issues are important. For instance, an absence of seed capital might inhibit new modern firms from starting up. The arrival of new information and communication technologies can, however, radically decrease the costs of sharing information to distant farms. Finally, all the standard information-related market failures related to the appropriation of knowledge that have given rise to research and development subsidies and tax incentives, and to intellectual property rights systems, also apply. Together with the incentives to accumulate, these constitute the enabling or operating environment that the various players in the agricultural space need to work. However, the challenge for an individual country is to identify where the most binding distortions or constraints lie and then remedy them. The fact that progress may have to be made on several fronts at once presents "the productivity policy dilemma": moving back from the technological frontier, governments face more market failures and distortions while their capabilities in remedying them become weaker (Fuglie et al. 2020).

REFERENCES

Fuglie, K., M. Gautam, A. Goyal, and W. F. Maloney. 2020. *Harvesting Prosperity: Technology and Productivity Growth in Agriculture.* Washington, DC: World Bank.

Townsend, R., L. Ronchi, C. I. Brett, and E. Moses. 2018. "Future of Food: Maximizing Finance for Development in Agricultural Value Chains." Washington, DC: World Bank Group. http://documents.worldbank.org/curated/en/593641523880972785/Future-of-Food-Maximizing-Finance-for-Development-in-Agricultural-Value-Chains.

APPENDIX D

Main Agricultural Innovations of the Past

This appendix provides information complementary to that in chapter 4.

GREEN REVOLUTION

The modern varieties (MVs) were by far the most successful innovations, developed initially by international agricultural research centers in partnership with national agricultural research systems. The first MVs were limited to cereal grains grown in irrigated areas or in areas with adequate rainfall. Subsequent agricultural research and development has extended improved technologies to more environments (rainfed or unsure irrigation) and more crops (such as root crops, pulses, other cereals) (Evenson and Gollin 2003; Hazell 2009). To reduce yield losses, in the mid-1970s the breeding programs also began developing MVs resistant to various pests and diseases (Khush 1987, 1989). In the late Green Revolution period (1981–2000), production gains were more dependent on MVs than in the early period, accounting for almost 50 percent of yield growth and 40 percent of production growth for all developing countries (Evenson and Gollin 2003). The widespread adoption of MVs, fertilizer, and pesticide also gave rise to innovation in machinery (box D.1).

Fertilizer and pesticide use have been rampant in East Asia. The adoption of MVs was already widespread by 2000, and both fertilizer and pesticides have continued increasing at high rates. Box D.2 discusses fertilizer and pesticide use in East Asia and compares it with other regions.

AGRICULTURAL INNOVATIONS FOSTERING PRODUCTIVITY AND SUSTAINABILITY

Both international agricultural research centers and national agricultural research systems have striven to develop MVs that are less reliant on pesticides and fertilizer, as well as to develop diverse sustainable cropping and livestock systems. Box D.3 offers details on a range of practices that are typically incorporated into environmentally sustainable commodity production systems.

Sustainable crop and livestock practices. A wide spectrum of innovations have been pursued with the goal of improving and maintaining productivity, sustainability, and health. Table D.1 summarizes the main innovations promoted to achieve the triple win.

BOX D.1

Innovation in the agricultural machinery industry in Asia

The 1980s and 1990s saw rapid expansion in the availability and use of low-cost, small-scale irrigation pump sets in Asia. As Green Revolution technologies made irrigation more profitable, farm demand for pump sets increased. Reforms in China allowed for-profit firms to manufacture and export pump sets while trade policy reforms in South and Southeast Asia reduced import restrictions on farm machinery. Hundreds of small and medium firms emerged in China to meet the growing market demand for small, low-cost pump sets among smallholder farmers in Asia, and the clustering of these firms in certain locations facilitated the spread of design innovations and standards across firms. More recently, rising rural wages in many Asian countries have led to greater demand for tractors and many specialized types of agricultural machinery. Policies can have important but complex influences on how private industry is able to respond to new market demands.

Source: Huang, Rozelle, and Hu 2007.

BOX D.2

Fertilizer and pesticide use in East Asia

Fertilizer use. Among regions, East Asia was by far the heaviest consumer of fertilizer in 2014, with annual usage averaging about 470 kilograms of nitrogen (N), phosphate (P205), and potash (K20) nutrients per hectare of arable and permanent cropland. This is in comparison to 140–145 kilograms in the European Union and South Asia; 115–125 kilograms in South America, North America, and Southeastern Asia; 70–85 kilograms in Western Asia, Central America, Australia, and New Zealand; 30–45 kilograms in Eastern Europe, the Caribbean, and Central Asia; and slightly more than 22 kilograms in Africa.

Within the East Asia region and across all crops, China and Malaysia had the highest use of fertilizer (as of 2014), followed by Vietnam and Indonesia. In comparison, Cambodia, the Lao People's Democratic Republic, Myanmar, the Philippines, and Thailand used much less fertilizer. Approximately two-thirds of the fertilizer consumed in Vietnam is used for rice (at about 180 kilograms of nutrients per hectare of harvested paddy in 2010–11), and other significant uses are for maize, coffee, and rubber. After experiencing steep growth, fertilizer consumption now seems to be stabilizing in China and Vietnam and falling in the Philippines, though accelerating for certain crops. Between 1961 and 2013, total nitrogen, phosphate, and potash fertilizer consumption rose 52-fold in China, 38-fold in Vietnam, and 8.5-fold in the Philippines (based on the International Fertilizer Association's IFADATA). In Vietnam, fertilizer application rates grew rapidly during the 1990s, but they have more or less stabilized since the mid-2000s. Fertilizer use has also flattened in China since late in the first decade of the 2000s, although the fact that it has flattened overall masks its rapid rise in specialty crop production. In the Philippines, where input use was lower to begin with, both fertilizer and pesticide use significantly declined between 2004 and 2013.

Pesticide use soared across much of China during the 1990s and 2000s and parts of Vietnam during the 2000s and 2010s, while also growing sharply but to far more moderate levels in the Philippines. China is the world's largest producer and consumer of pesticides, as well as one of its more intensive users of these chemicals on a per hectare basis. It used more than 1.8 million tons of active ingredients in 2014 (a 136 percent increase over 1991 levels, or a 7.5 percent increase per year), and an estimated 14.8 kilograms per hectare as of 2012, or more than two-and-a-half times the 1991 intensity (based on FAOSTAT data). Vegetables, as a group, accounted for the largest sales of pesticides in China as of 2006, at 24 percent, and rice was the single

continued

Box D.2, *continued*

largest user of pesticides, at 15 percent (Zhang, Jiang, and Ou 2011). China's average pesticide consumption was reportedly two-and-a-half times the world average and was still increasing as of 2014, although its growth has slowed in the 2010s, and a number of toxic chemicals have been banned or are being phased out (Gao et al. 2017).

Within the focus countries in East Asia, Malaysia is the next most intensive user of pesticides after China. Vietnam's use of pesticides is more moderate, but, as with fertilizer, its geographic concentration in major crop production areas makes it a concern, and changes in cropping patterns and pest resistance may have since driven more intensive use. In the Philippines, overall pesticide use is not only lower, but recently has been falling. Nevertheless, the use of pesticides in the country's large-scale, commercial fruit plantations is shrouded in statistical opacity and is abundant by some accounts (Havemann and Rosenthal 2015; Magcale-Macandog et al. 2016).

Source: Cassou, Jaffee, and Ru 2017.

BOX D.3

Practices to mitigate environmental risks

Some common practices and examples of environmentally sustainable practices include the following:

- *Agricultural soil restoration and rehabilitation.* Soil organic matter enrichment, crop rotation, appropriate tillage and management practices
- *Soil conservation.* Plot-level soil management, appropriate siting of crop production within the farm or landscape, vegetative cover on farmed and nonfarmed land, vegetative and earth barriers to soil movement
- *Improved management of surface or groundwater.* On-site water and soil management, re-vegetation, investment in water harvesting structures
- *Improved management of irrigation and stream flow.* Well-designed drainage systems, flood control structures, strategic siting of infrastructure, controlled diversion of water for irrigation and industrial or urban use
- *Improved agrochemical input management.* Best practices in agrochemical application, vegetative and structural barriers
- *Permanent vegetative cover.* Using cover crops; retaining areas of natural perennial grasses, trees, or palms; intercropping or retaining productive trees and shrubs in cropland
- *Native habitat restoration.* Farm and community habitat networks and habitat set-asides, protected area establishment and restoration, connectivity between communities and protected areas
- *Wildlife protection in and around production areas.* Maintaining wildlife access to uncontaminated water and food sources, enabling mobility across farm fields; hunting and gathering at sustainable levels; using integrated pest management systems
- *Diversification of land cover.* Multicropping, using multiple crop varieties or cultivars; farm diversification; community or landscape diversification
- *Greenhouse gas mitigation.* Asia has the highest total technical mitigation potential in the world, via carbon sequestration through enhanced cropland and grazing land management, restoring organic soils and degraded lands, better practices in rice management, and reduced deforestation and degradation (Smith et al. 2014).

Sources: McNeely and Scherr 2003; Scherr et al. 2015; Smith et al. 2014; WOCAT 2007.

TABLE D.1 **Current triple win innovations in use or in the early stages of adoption in East Asia**

INNOVATION PRACTICE[a]	ADOPTION	CHANGES IN PRODUCTIVITY, ENVIRONMENT, OR HEALTH	ECONOMIC FEASIBILITY AND EQUITY ASPECTS
	Overall crop-livestock management		
System for Rice Intensification (SRI)—also a CSA	Most countries have pilots Vietnam: More than 1 million farmers (by 2011), or 10% of farmers China (South): A mainstream practice	Increased yield and profits, savings in inputs (seed, water), improved soil condition and plant health China yield increase of 45–100%	Feasible despite knowledge-intensiveness
CSA in rice China: Fixed-film rice production in hillier colder regions, saturated fields and a plastic film cover; low methane cultivars	30–60% adoption rate in Vietnam < 30% in the Philippines China (fixed film rice): Widespread in cooler and hillier areas China (low methane cultivars): Limited research so far	Emissions down 50% Potential exists	Feasible
Landscape approaches	Mostly pilot stage (Cambodia, Vietnam), limited adoption	Potential for all, but limited economic benefits	Complex—requires coordination across communes and regions
CSA in livestock Vietnam (biogas systems, local feed) Philippines (biogas, composting, feed) China, Hubei province offers livestock farmers concessional loans for investments in covered manure sheds, biogas digesters, and other manure management upgrades	Vietnam < 30%; 30–60% Philippines < 30%; 30–60% China—not available (Cassou, Jaffee, and Ru 2017)	Not yet known	High up-front investment in biogas systems may prevent adoption; may consider incentives for and new economic use for (manure) adoption
CSA in shrimp	Vietnam: Shrimp production in coastal areas: shrimp-rice, shrimp-tilapia, or shrimp-forest farming with mangrove systems	—	Feasible
CSA in wheat and corn Wheat: Adjusting sowing date, dense planting, and improving grain filling Corn: Adjusting sowing date, dense planting, and mulching	China, becoming mainstream practices	Yield maintained, reduction in costs	Feasible
Input management programs on rice Vietnam, Mekong Delta: 1M5R program Vietnam, South: 3R3G in rice program (reduced use of seeds, nitrogen fertilizer, and pesticides) A "small farmer, large field" model in Vietnam	3R3G : Adopted reduced use of seed (Huelgas et al. 2008) China: Agro-input precision use	1M5R: Rice yields maintained while costs per hectare reduced 18–25% (Nguyen et al. 2015)	Feasible

continued

TABLE D.1, *continued*

INNOVATION PRACTICE[a]	ADOPTION	CHANGES IN PRODUCTIVITY, ENVIRONMENT, OR HEALTH	ECONOMIC FEASIBILITY AND EQUITY ASPECTS
Integrated production of coffee	Coffee in Vietnam's Central Highlands: Use of fertilizer and water (Amarasinghe et al. 2015; TechnoServe 2013)	Yield increase, input savings	Feasible
	Soil fertility and structure		
Soil testing kits, IRRI nitrate tools	Testing in many locations but adoption rate varies due to availability and cost		Feasible, yet test cost may discourage some farmers
Improved fertilizer formulas (granules, slow release), timing and dosing practices, manure management practices	China: Hebei province has started subsidizing less-polluting fertilizer that releases nutrients gradually and is applied based on soil testing and crop needs (Cassou, Jaffee, and Ru 2017)	n.a.	Feasible if availability of formula fertilizers improves
	China: Pilots in use of nitrogen inhibitors (keeps nitrogen stable) and better use of manure (fields rather than rivers) and better manure management (dry rather than wet) (Searchinger 2016)		Investment in application tools and availability and cost of service providers an obstacle Current fertilizer subsidies (industry, farmers) are counterproductive
Biofertilizers	Most countries apply biofertilizers in both conventional and organic agriculture. Data on use and application not available. China: Rhizobium extensively on many crops (rice, wheat) Vietnam: *Burkholderia vietnamiensis* TVV75 and *Pseudomonas aeruginosa* (rice, watermelon)	n.a.	Feasible: Inexpensive
Integrated nutrient management Integrated soil-crop management practices able to compensate for additional nitrogen fertilizer application	China: Data on adoption not available	Yield increase from 7.2 metric tons per hectare to 8.5 metric tons per hectare	Feasible
Erosion and soil fertility management China Loess Plateau Vietnam export crops: Reducing soil erosion in mountainous areas (for example, via contour farming in maize, planting grass strips along contour lines (Mulato, Guinea) and improving soil fertility by intercropping with leguminous species (cassava or rubber)	China: Soil erosion control and productivity Philippines: Widespread use of, for example, velvet in soil erosion Vietnam export crops: 30–60%	Yield and soil improvement, erosion control — —	Land tenure rights may limit farmers' willingness to commit to long-term investment in soil erosion approaches

continued

TABLE D.1, *continued*

INNOVATION PRACTICE[a]	ADOPTION	CHANGES IN PRODUCTIVITY, ENVIRONMENT, OR HEALTH	ECONOMIC FEASIBILITY AND EQUITY ASPECTS
	Pest management		
Integrated pest management	Limited adoption although promoted for decades in all East Asian countries	Reduced pesticide use, no effect on yield, savings in costs	Effective approach, but as a knowledge-intensive approach it is expensive to spread among smallholders
Biopesticides	Several countries in the region have adopted biopesticides, with biopesticides based on *Bacillus thuringiensis* (Bt) being most widely used China accounts for 35% of the overall market and is expected to be the fastest growing in the region because of increasing acceptance of biopesticides as an alternative to existing chemical pesticides	Positive health effects, slow effect on pests	Feasible
Smaller or redesigned chemical containers, painted pesticides (on seed)	China: No data available	—	Feasible
GM cotton	China (90–100%), Myanmar (92%), Malaysia, Indonesia (coming to market), Philippines	Improved yield, reduced pesticide and labor, fewer health incidents	Cost-effective, yet seed cost and availability may prevent adoption
GM maize	Vietnam 4% (2017); Philippines 44% of corn (2017); Indonesia, Malaysia, China	—	
GM soybean	Vietnam, Indonesia, Malaysia, Philippines	Not yet known	
GM sugarcane	Indonesia (coming to market)		
GM papaya	China, widespread		
GM potato	Malaysia, Philippines		
GM oilseed rape	China, Malaysia, Philippines		
Water management			
Drip irrigation and micro dosing	All countries, no adoption data available	—	Feasible for special crops; up-front cost
Water use efficiency practices	Vietnam (coffee): 30–60% adoption	—	—
Alternate wetting and drying	China: Widespread in selected areas, otherwise limited	Water saving 40%, emissions down 20–90%, yield decrease of 5.4%	Feasible
	Other		
Biofortified crops: Vitamin A orange sweet potato, iron beans, iron pearl millet, vitamin A yellow cassava, vitamin A orange maize, zinc rice, and zinc wheat	142 countries—HarvestPlus, a CGIAR research program on Agriculture for Nutrition and Health, focuses on three micronutrients: iron, zinc, and vitamin A (https://www.harvestplus.org/what-we-do/crops)	Productivity and health	—

continued

TABLE D.1, *continued*

INNOVATION PRACTICE[a]	ADOPTION	CHANGES IN PRODUCTIVITY, ENVIRONMENT, OR HEALTH	ECONOMIC FEASIBILITY AND EQUITY ASPECTS
	Other		
GM golden rice (vitamin A-rich)	Philippines from 2020 onward	Potential to improve health	Not yet known
GM potato (quality aspects)	Malaysia	Not yet known	
GM soybean (high stearidonic acid, lower level fatty acids, increased oleic acid)	Indonesia, Philippines		
GM maize (drought tolerance and thermostable alpha-amylase production)	Indonesia, Philippines		

Sources: Original table for this publication. Amarasinghe et al. 2015; Cassou, Jaffee, and Ru 2017; Huelgas, Templeton, and Castanar 2008; Nguyen et al. 2015; Searchinger 2016; TechnoServe 2013. GM data from FAO GM Foods Platform, accessed February 2020 (http://www.fao.org/food-safety), and ISAAA 2019.
Note: — = not available; 1M5R = 1 Must 5 Reductions program; 3R3G = Three Reductions Three Gains program; CSA = climate-smart agriculture; GM = genetically modified; IRRI = International Rice Research Institute; n.a. = not applicable; SRI = system for rice intensification.
a. Innovation practice refers to a wide range of innovations that have been promoted to improve some combination of productivity, environmental sustainability, and health.

Rice production practices. Rice production in most East Asian countries is associated with significant environmental challenges and is also the biggest generator of greenhouse gas (GHG) emissions. Box D.4 describes some of the sustainable production practices promoted in rice production. Box D.5 describes evidence from precision application of inputs in China.

Improvements in water management, along with technical innovations, are an important part of future crop production. Box D.6 discusses good practices in water users association management.

Integrated livestock practices that can lower the environmental footprint and GHG emissions of livestock production rely on multiple strategies (action area) and innovations (table D.2). Some of the practices have been well tested and under implementation (best practices), whereas others are at the experimental or early commercialization stage.

Climate-smart agriculture. The climate-smart agriculture (CSA) sector covers a broad range of innovations, services, and approaches that help farmers and the sector to adapt to and mitigate climate change effects (chapter 4) (Dikitanan et al. 2017; Fuglie et al. 2020; Green 2018; Nguyen et al. 2017). However, countries vary in their readiness to promote and adopt CSA. For instance, Myanmar is one of the most vulnerable countries in the region, yet it is in the very early stages of CSA adoption[1] (World Bank 2019), while the more affluent China, the Philippines, and Vietnam have taken steps forward in pursuing CSA (box D.7). Vietnam is among the countries most vulnerable to climate change—climate change is expected to reduce the agricultural production area by about 12 percent in the Red River Delta and 24 percent in the Mekong River Delta. Climate change will likely affect not only the agricultural production area but also agriculture productivity. CSA practices have been identified in Vietnam, and their adoption rate ranges mostly between 30 percent and 60 percent. The agricultural research and development system, along with extension, is actively promoting diverse

BOX D.4

Sustainable rice production practices

A broad array of production, natural resource, and farm management techniques can help reduce or eliminate the adverse environmental impacts of rice. The Sustainable Rice Platform brings many stakeholders together to develop guidelines and strategies that will be broadly applicable.

Input use efficiency. Initiatives promoting input use efficiency have had major benefits for farmers. In 1986, the president of Indonesia abolished subsidies for pesticide use in rice production and established an integrated pest management (IPM) program. These changes led to fewer pest outbreaks, higher rice yields, less money spent by farmers on pesticides, and more than $1 billion saved by the Indonesian government over 10 years. Although IPM has faced challenges in scaling up (small farm size, knowledge-intensive approach), the approach has empowered some farmers to reduce the use of pesticides, especially the most toxic ones, by using them as a last resort and favoring reliance on prevention, biological controls, and, when necessary, lower-toxicity, lower-residue, and high-efficiency pesticides. IPM has also contributed to reduced pesticide use in the Philippines.

Nutrient management tools. In various parts of the region, nutrient management tools, including ones that bypass soil testing, have proven effective at reducing fertilizer use, along with waste and imbalances. An example of this is the questionnaire-based Rice Nutrient Manager tool developed by the International Rice Research Institute for use by farmers and extension workers via mobile devices (Buresh et al. 2012).

Alternate wetting and drying (AWD). Rice requires improving water productivity, and water-saving irrigation technologies have been adopted, such as AWD irrigation, which has widely been implemented in many areas in China. In AWD, alternating flood and nonflood conditions are practiced in the field. The depth of standing water will gradually decrease after irrigation, and when it drops 15 centimeters below the surface of the soil, the field is irrigated to a depth of approximately 5 centimeters. AWD has been shown to reduce water requirements by up to 40 percent and greenhouse gas emissions by 20–90 percent (Lampayan et al. 2015; Linquist, Snyder, and Anderson 2015). However, according to a recent meta-analysis, AWD adoption remains limited because of its negative effects on rice yield (Carrijo, Lundy, and Linquist 2016).

Climate-smart agriculture (CSA) practices. In Vietnam, the main two CSA practices for rice include a system of rice intensification (adoption rate less than 30 percent) and use of flood-tolerant rice varieties (adoption rate 30–60 percent). Changes in fertilizer use, together with AWD (a technique that uses less water than the more conventional permanent flooding), have in some instances more than halved emissions of carbon dioxide equivalent while saving water and improving yields (Thu et al. 2016). Methane emissions from rice paddies can also be lessened using improved rice cultivars and fertilization techniques, including ones that use rice straw, thereby averting burning (Adhya et al. 2014; Thu et al. 2016). In the Philippines, given the different growing conditions, CSA for rice covers a different set of practices, such as water harvesting and use of site-specific nutrient management and pest management practices. In China, the CSA in rice production concerns adjustment of cropping regions, dense planting, use of non-flooded irrigation, and use of superior rice cultivars. For years China has pursued breeding of super rice varieties and has adopted improved nursery practices and use of AWD in rice fields (similar to SRI). Hillier and colder regions in China have also adopted so-called fixed-film rice production, that is, maintenance of saturated—not flooded—fields covered with a plastic film, which can reduce more than half of emissions.

Landscape initiatives. Landscape-scale rice initiatives are less common. In one Cambodian example, farmers living on or adjacent to protected areas received higher rice prices in exchange for protecting species such as the endangered giant ibis. Farmers who agreed to limit their hunting and forest clearing practices were able to sell their output through the Wildlife Friendly™ brand. The project succeeded in decreasing habitat clearing by 50 percent among participating farmers (Nielsen, Ashish, and Clements 2015). In the Philippines, small-scale communal irrigation schemes created better-paying jobs that

continued

Box D.4, *continued*

reduced incentives to clear forest and extract products in upland areas. Annual forest clearing by upland households adjacent to the lowland irrigated area declined by 48 percent (McNeely and Scherr 2003).

In Vietnam's Mekong Delta region, efforts to reduce greenhouse gas emissions associated with rice are being increased through landscape-scale initiatives that also increase input use efficiency. Thus, collaboration among stakeholders such as farmers, conservationists or governments, and private companies can certainly lead to greener, more productive rice systems.

Additional sources: Abraham et al. 2014; Cassou, Jaffee, and Ru 2017; Chen et al. 2014; Dzung 2012; Huelgas, Templeton, and Castanar 2008.

BOX D.5

Precision application of inputs: Innovations and challenges with smallholders in China

Over the past decade, China's Ministry of Agriculture has actively promoted several technologies to help farmers make more judicious use of chemical inputs. For example, since the early 2000s the Ministry of Agriculture has financed soil testing and promoted the diffusion of formula fertilizer based on soil testing results. In theory, this technology allows farmers to use tailor-mixed fertilizers that better align with their specific needs, considering local conditions and what they are growing. A World Bank project has promoted this technology in Guangdong Province by competitively procuring fertilizers specially formulated based on soil testing and crop needs and creating incentives for farmers to use them. After three years of efforts to promote formula fertilizer, the market seems to have responded positively. As of 2017, multiple brands of formula fertilizer that have *not* been procured under the project have become available in agro-input shops in project areas.

Formula fertilizer is not a panacea, however. Various circumstances have limited the technology's effectiveness in pollution control. Government-sponsored soil testing is still limited (at one sample per 10–20 hectares), and soil testing results are not systematically made available to input suppliers. Furthermore, tailor formulating fertilizers to meet a wide range of smallholder needs remains cost prohibitive for most suppliers. Meanwhile, formula fertilizer does not address the issue of poor timing, a problem that is growing as increasing numbers of farmers seek off-farm employment.

Although new technologies and service models hold promise for addressing this major challenge, those that have emerged to date have a long way to go before they reach commercial viability. Since 2007, for example, the market for slow- and controlled-release fertilizers has been growing in China (Heffer 2016). In Guangdong Province, however, adoption of such fertilizers is still limited because of concerns about cost and quality. One limitation of this technology is its current inability to provide a boost of nutrients as needed in a crop's growth cycle.

In parallel, the development of professional pesticide application services has also faced challenges. Farmers often prefer to manage pests on their own when they do not have off-farm jobs. The development of pesticide application services has also been complicated by disputes over responsibility for pest outbreaks and partial participation by farmers in a given area. Partial participation on the part of farmers undermines the quality of the service because a pest infestation in a nonparticipating plot can undermine the efficacy of the pesticide application service in the surrounding area.

Source: Cassou, Jaffee, and Ru 2017.

BOX D.6

The impact of water users associations on farm production, income, and water savings in northern China

Studies have examined the impact of water users associations (WUAs) on farm production, income, and water savings (Wang et al. 2005; Wang et al. 2006; Wang et al. 2010). These studies find that WUAs have not been universally successful in either saving water or improving farm incomes and link the performance of water management systems to the incentives that these new institutions provide to water managers. Wang et al. (2010) identify five key principles that, according to World Bank project managers, WUAs should satisfy to be successful: (1) adequate and reliable water supply; (2) the WUA should be organized hydraulically (not administratively); (3) leaders should be elected and WUA management and decision-making should be with the farmers (without local government interference); (4) water should be charged volumetrically (not according to land area); and (5) the WUA should have the right to collect water fees.

Empirical evidence among WUAs in Ningxia, Gansu, Hubei, and Hunan provinces indicates that there are important differences in the extent to which these five key principles are implemented, and that the degree of implementation has significant implications for water use efficiency (Wang et al. 2010). Water use in rice, wheat, and maize in World Bank–supported WUAs, which mostly operate according to the five principles, is found to be 15–20 percent lower than in traditionally managed villages. In villages where participation by farmers plays only a minor role and water management reforms have been only nominally implemented, the establishment of WUAs has had little effect on water use. The study further finds that crop yields and incomes are not significantly different between World Bank–supported WUAs and other WUAs.

Zhang et al. (2013) also find that group characteristics, particularly group size and number of water users groups, and the existing pressure on available water resources are important factors in water productivity. Large groups tend to have greater difficulties in overcoming problems of collective action and free-riding. A large number of subgroups, that is, water users groups, within a WUA can promote water productivity by allowing more crop diversification and by better tuning planting and irrigation decisions among member households. Another group characteristic that affects water productivity in the sample is heterogeneity of land endowments, which was found to have a positive effect on water productivity of member households in a WUA. WUAs with a relatively small number of member households, a large number of water users groups, and low pressure on the available water resources were more likely to achieve relatively high water use efficiencies.

adaptation and input efficiency programs for rice and selected commodities. The assessment is that Vietnam is ahead of its regional neighbors in CSA for rice. The overall agriculture sector of the Philippines is, however, quite different and underdeveloped, as reflected in the low uptake of CSA practices, which is mostly a result of poor availability of and access to improved seed, insufficient financial resources to cover investment costs, and constrained extension services.

China is also taking strides in the CSA area.[2] China is by far the dominant emitter of agricultural GHGs in the region. Given the nature of agricultural production in China, reducing GHG emissions from fertilizer use shows the most potential among CSA practices (Searchinger 2016). Application of new fertilizer formulas (granules, slow release, with nitrogen inhibitors), applications based on soil testing and appropriate timing and dosing, better manure management (dry), and water management in rice fields (frequency of drainage, saturated field, and plastic film) have proven appropriate—yet adoption is still limited. In the livestock sector (the greatest emitting subsector) ongoing research

TABLE D.2 **Practices for improving sustainable and low-carbon livestock production**

ACTION AREA	BEST PRACTICES AND PILOTS FOR COMMERCIALIZATION	PROOF OF CONCEPT AND DISCOVERY STAGE
Feed and nutrition	Improving forage quality and digestibility Dietary improvements and substitutes Precision feeding, feed supplements (for example, probiotics, nucleotides, enzymes)	Active innovation area
Animal genetics and breeding	Efficient and robust animals Improved performance on low-quality feed	Selecting for low-methane-producing ruminants Finding new traits for greenhouse gas emissions
Rumen modification	—	Inhibitors, transferring the microbiome of low-methane-producing ruminants Vaccines to reduce methane production
Animal health	Increasing productive lifetime of animals Prevention, control, and eradication of disease Probiotic feed additives for reducing antimicrobial overuse	Increase disease resistance
Manure management	Collection and storage facility Temperature and aeration of manure Capturing biogas from anaerobic process Manure deposition and application, storage cover	—
Grassland management	Grazing practices Pasture management	Carbon sequestration

Sources: Adapted from FAO 2019; GRA and SAI 2013; Mitchell 2019.
Note: — = not available.

BOX D.7

Climate-smart agriculture practices for key crops and commodities in the Philippines and Vietnam

Agriculture is the second-largest source of national greenhouse gas emissions (including short- and long-lived greenhouse gases) in the Philippines and Vietnam, making it of key importance to meeting commitments to mitigate climate change (Cassou, Jaffee, and Ru 2017). Practices are considered climate smart if they enhance food security as well as at least one of the other objectives of climate-smart agriculture (CSA)—adaptation or mitigation. Tables 4.4 and 4.5 in chapter 4 present the main CSA practices and their adoption rates in Vietnam and the Philippines. The CSA practices in Vietnam include the following key groups.

Smart water and irrigation management in almost all crop systems, that is, coffee, tea, citrus, cashew, maize, rice, and pepper. This practice refers to installing water-saving irrigation techniques such as drip or sprinkler irrigation, implementing moisture-preserving practices such as mulching (cassava), alternate wetting and drying systems in rice (a component of the system of rice intensification technique), input-saving techniques in rice production (1 Must 5 Reductions program, Three Reductions Three Gains program), integrating fishponds into citrus plantations, and use of humus storage pits in rubber.

Adoption of improved crop varieties resistant to droughts, floods, or pests and diseases (for example, in rubber, cashew, cereals, and pepper), which can also support *integrated pest management*. Integrating perennials such as orange, rubber, coffee, or cashew into *agroforestry systems* with other crops such as avocado, maize, guava, upland rice, or sesame would allow

continued

Box D.7, *continued*

farmers to diversify their incomes and improve their productivity and resilience to climate change. Intercropping also allows for *heat stress management (microclimate regulation)* in coffee by planting shade trees such as durian fruit trees or reduces direct soil moisture loss by intercropping with leguminous plants (peanut, *Arachis pintoi*, and others).

Sustainable land management reduces soil erosion in mountainous areas (for example, via contour farming in maize or planting grass strips along contour lines [Mulato grass or Guinea grass]) and improves soil fertility by intercropping with leguminous species, for example, in cassava or rubber.

Manure management and improved feed and fodder management. For livestock, the practices suggest the integration of biogas technologies into pig production for *efficient manure management,* and *improved feed and fodder management* such as the use of local high-quality feeds that are more easily available and increase productivity. Shrimp production is practiced by farmers in coastal areas as a means to deal with saltwater intrusion, and can be integrated into shrimp-rice, shrimp-tilapia, or shrimp-forest farming with mangrove systems for higher productivity.

The agriculture sector of the Philippines is somewhat different from that of Vietnam. A combination of farm characteristics (small scale and fragmented), a lack of infrastructure, and policy and institutional barriers has left the sector underdeveloped and unable to meet the food requirements of the growing population. Crops occupy roughly 76 percent of the harvested land, with rice, coconut, and maize occupying the largest share of this area. Because of its geographic location and archipelagic formation, the Philippines is one of the region's most vulnerable to the impacts of climate change. The country ranks highest in the world in vulnerability to tropical cyclones, third in people exposed to such seasonal events, and fourth among countries most affected by extreme weather events. As a result of increased water and heat stress, climate change and variability are expected to decrease crop yields, increase the incidence of pests and diseases, and cause shifts in crop production suitability. Upland areas, for example, will potentially benefit from increases in temperature by creating adequate conditions for the growth of new crops.

Sources: Cassou, Jaffee, and Ru 2017; Dikitanan et al. 2017; Nguyen et al. 2017.

focuses on reducing methane emissions by adding various additives to animal diets (probiotics, lipids, tea saponins) and, in some circumstances, methane emissions from ruminant animals have been decreased (Powlson, Norse, and Lu 2018). In wheat production, the main practices concern adjusting sowing dates, dense planting, and improving grain filling. In rice production, the main practices include adjusting cropping regions, dense planting, use of nonflooded irrigation, and use of superior rice cultivars. For years China has pursued the breeding of super rice varieties and has adopted improved nursery practices and the use alternate wetting and drying. In corn production, the main practices include adjusting sowing dates, dense planting, and mulching.

NOTES

1. Myanmar's government allocated most public expenditures for rice cultivation, whereas the sustainable management of livestock and manure treatment received very little in the government budget. Furthermore, the large investments in irrigation and flood control were not climate smart, given that they did little to strengthen on-farm irrigation infrastructure and were not complemented by programs such as alternate wetting and drying and soil nutrient management (World Bank 2017).

2. Overall, China's efforts focus on both adaptation and mitigation. The key adaptation strategy is to (1) enhance the resilience of agriculture through improved irrigation, (plastic) mulching, and good planning of cropping cycles; and (2) implement conservation farming through soil quality enhancement with organic matter and crop straw incorporation. The key mitigation strategy is to reduce nitrous oxide emissions through enhanced nitrogen use efficiency, including formula fertilization, soil testing, slow-release fertilizer, and organic fertilizer.

REFERENCES

Abraham, B., O. O. AdeOluwa, H. Araya, T. Berhe, Y. Bhatt, S. Edwards, B. Gujja, et al. 2014. *The System of Crop Intensification: Agroecological Innovations for Improving Agricultural Production, Food Security, and Resilience to Climate Change*. Ithaca: SRI International Network and Resources Center and Technical Centre for Agricultural and Rural Cooperation.

Adhya, T. K., B. Linquist, T. Searchinger, R. Wassmann, and X. Yan. 2014. "Wetting and Drying: Reducing Greenhouse Gas Emissions and Saving Water from Rice Production." Working Paper 8 of "Creating a Sustainable Food Future," World Resources Institute, Washington, DC. http://www.wri.org/sites/default/files/wetting-drying-reducing-greenhouse-gas-emissions-saving-water-rice-production.pdf.

Amarasinghe, U. A., C. T. Hoanh, D. D'haeze, and T. Q. Hung. 2015. "Toward Sustainable Coffee Production in Vietnam: More Coffee with Less Water." *Agricultural Systems* 136: 96–105. http://dx.doi.org/10.1016/j.agsy.2015.02.008.

Buresh, R. J., R. Castillo, M. van den Berg, and G. Gabinete. 2012. *Nutrient Management Decision Tool for Small-Scale Rice and Maize Farmers*. Taipei: Food and Fertilizer Technology Center. http://www.agnet.org/library.php?func=view&id=20140305111316.

Carrijo, R. D., M. Lundy, and B. A. Linquist. 2016. "Rice Yields and Water Use under Alternate Wetting and Drying Irrigation: A Meta-Analysis." *Field Crops Research* 203 (1): 173–80. https://doi.org/10.1016/j.fcr.2016.12.002.

Cassou, E., S. M. Jaffee, and J. Ru. 2017. *The Challenge of Agricultural Pollution: Evidence from China, Vietnam, and the Philippines*. Directions in Development. Washington, DC: World Bank.

Chen, X., Z. Cui, M. Fan, P. Vitousek, M. Zhao, W. Ma, Z. Wang, et al. 2014. "Producing More Grain with Lower Environmental Costs." *Nature* 514 (7523): 486–89. doi:10.1038/nature13609.

Dikitanan, R., G. Grosjean, A. Nowak, and J. Leyte. 2017. "Climate-Resilient Agriculture in Philippines." CSA Country Profiles for Asia Series. International Center for Tropical Agriculture (CIAT); Department of Agriculture—Adaptation and Mitigation Initiatives in Agriculture, Government of the Philippines, Manila.

Dzung, N. 2012. "Simple and Effective SRI and Agriculture Innovation." SRI International Network and Resources Center, Ithaca.

Evenson, R. E., and D. Gollin, eds. 2003. *Crop Variety Improvement and Its Effect on Productivity: The Impact of International Agricultural Research*. Wallingford, UK: CABI.

FAO (Food and Agriculture Organization of the United Nations). 2019. *Five Practical Actions towards Low-Carbon Livestock*. Rome: FAO.

Fuglie, K., M. Gautam, A. Goyal, and W. F. Maloney. 2020. *Harvesting Prosperity: Technology and Productivity Growth in Agriculture*. Washington, DC: World Bank.

Gao, S., Q. Wang, D. Guan, W. F. Zhang, Y. Li, Z. Shi, C. Yan, et al. 2017. "An Overview of Agricultural Pollution in China." Unpublished paper prepared for the World Bank, Washington, DC.

GRA and SAI (Global Research Alliance on Greenhouse Gases and Sustainable Agriculture Initiative). 2013. "Reducing Greenhouse Gas Emissions from Livestock: Best Practice and Emerging Options." GRA and SAI. https://saiplatform.org/uploads/Modules/Library/lrg-sai-livestock-mitigation_web2.pdf.

Green, Adam. 2018. "Food 4.0: The Future of Food Innovation in Asia." Economist Intelligence Unit Limited.

Havemann, T., and N. Rosenthal. 2015. "Banana Production in Mindanao, Philippines." In *Steps toward Green: Policy Responses to the Environmental Footprint of Commodity Agriculture in East and Southeast Asia*, edited by S. Scherr, K. Mankad, S. Jaffee, and C. Negra. Washington, DC: EcoAgriculture Partners.

Hazell, P. B. R. 2009. "The Asian Green Revolution." Discussion Paper 00911, International Food Policy Research Institute, Washington, DC.

Heffer, P. 2016. "Fertilizer Consumption Trends in China vs. the Rest of the World." Presentation, International Symposium on Improvement of Nutrient Use Efficiency in China under Zero Growth of Chemical Fertilizers, Beijing, March 17–18. http://china.ipni.net/ipniweb/region/china.nsf/0/CDB5F6A3E2B492DA48257F7F004FF54D/$FILE/NUE2016-hefferp%20-%20PPT.pdf.

Huang, Q., S. Rozelle, and D. Hu. 2007. "Pump-Set Clusters in China: Explaining the Organization of the Industry That Revolutionized Asian Agriculture." *Asia-Pacific Development Journal* 14 (December): 75–105.

Huelgas, Z. M., D. Templeton, and P. Castanar. 2008. "Three Reductions, Three Gains (3R3G) Technology in South Vietnam: Searching for Evidence of Economic Impact." Contributed paper at the 52nd Annual Conference of the Australian Agricultural Resource Economics Society, Australia, February 5–8.

ISAAA (International Service for the Acquisition of Agri-biotech Applications). 2019. Global Status of Commercialized Biotech/GM Crops: 2019. ISAAA Brief 55. https://www.isaaa.org/resources/publications/briefs/55/.

Khush, G. S. 1987. "Rice Breeding: Past, Present, and Future." *J. Genetics*, 66: 195–216.

Khush, G. S. 1989. "Multiple Disease and Insect Resistance for Increased Yield Stability in Rice." In *Progress in Irrigated Rice Research*. Los Banos, Philippines: International Rice Research Institute.

Lampayan, R. M., R. M. Rejesus, G. R. Singleton, B. A. M. Bouman. 2015. "Adoption and Economics of Alternate Wetting and Drying Water Management for Irrigated Lowland Rice." *Field Crops Research* 170 (January): 95–108.

Linquist, R., R. Snyder, and F. Anderson. 2015. "Water Balances and Evapotranspiration in Water- and Dry-Seeded Rice Systems." *Irrigation Science* 33 (2015): 375–85.

Magcale-Macandog, D. B., R. M. Briones, A. D. Calub, R. B. Saludes, M. L. A. Cuvin-Aralar, A. R. Salvacion, E. V. P. Tabing, et al. 2016. *An Overview of Agricultural Pollution in the Philippines: Summary Report 2016*. Washington, DC: World Bank.

McNeely, J. A., and S. J. Scherr. 2003. *Ecoagriculture: Strategies to Feed the World and Save Wild Biodiversity*. Washington, DC: Island Press.

Mitchell, M. 2019. "Innovation with Livestock Feed: The Future of Animal Source Foods." *Agrilinks* (blog), October 2, 2019. https://www.agrilinks.org/post/innovation-livestock-feed-future-animal-source-foods.

Nguyen, T. H., T. H. Tran, M. D. Chau, and V. S. Nguyen. 2015. "Monitoring and Evaluating of the Application of Low Carbon Rice Production Technologies 1M5Rs and Its Impacts in Kien Giang and Giang Provinces. Workshop Proceedings." Vietnam Low Carbon Rice Project (VLCRP), Project-End Dissemination and Regional Policy Dialogue Workshop, Kien Giang, April 15.

Nguyen, T. T. N., F. Roehrig, G. Grosjean, D. N. Tran, and T. M. Vu. 2017. "Climate Smart Agriculture in Vietnam." CSA Country Profiles for Asia Series, International Center for Tropical Agriculture (CIAT), Food and Agriculture Organization, Hanoi.

Nielsen, K., J. Ashish, and T. Clements. 2015. "Ibis Rice: Helping Cambodian Farmers and Protecting Wildlife." In *Shades of Green: Multistakeholder Initiatives to Reduce the Environmental Footprint of Commercial Agriculture*, edited by S. Mirvat and S. Jaffee. Washington, DC: EcoAgriculture Partners.

Powlson, D., D. Norse, and Y. Lu. 2018. "Agricultural Development in China: Environmental Impacts, Sustainability Issues and Policy Implications Assessed through China-UK Projects under SAIN (UK China Sustainable Agriculture Innovation Network), 2008–2017." SAIN Working Paper 1.

Scherr, Sara J., Kedar Mankad, Steven Jaffee, and Christine Negra, with Tanja Havemann, Janjarang Kijtikhun, Uray Endang Kusumjaya, Samiksha Nair, and Naomi Rosenthal. 2015. *Steps toward Green: Policy Responses to the Environmental Footprint of Commodity Agriculture in East and Southeast Asia.* Washington, DC: EcoAgriculture Partners and the World Bank.

Searchinger, T. 2016. "Possibilities for Modernization of Agriculture and Mitigation of Its Greenhouse Gas Emissions in China." Short Discussion Memo of Possible Areas of World Bank/China Collaboration. Draft—November 18.

Smith P., M. Bustamante, H. Ahammad, H. Clark, H. Dong, E. A. Elsiddig, H. Haberl, et al. 2014. "Agriculture, Forestry and Other Land Use (AFOLU)." In *Climate Change 2014: Mitigation of Climate Change.* Contribution of Working Group III to the Fifth Assessment Report of the Intergovernmental Panel on Climate Change, edited by O. Edenhofer, R. Pichs-Madruga, Y. Sokona, E. Farahani, S. Kadner, K. Seyboth, A. Adler, et al. Cambridge, UK, and New York: Cambridge University Press.

TechnoServe. 2013. "A Business Case for Sustainable Coffee Production—Vietnam: An Industry Study by TechnoServe for the Sustainable Coffee Program." Report prepared for IDH, the Sustainable Trade Initiative, Hanoi.

Thu, T. N., L. B. T. Phuong, T. M. Van, and S. N. Hong. 2016. "Effect of Water Regimes and Organic Matter Strategies on Mitigating Green House Gas Emission from Rice Cultivation and Co-benefits in Agriculture in Vietnam." *International Journal of Environmental Science and Development* 7 (2): 85. http://www.ijesd.org/vol7/746-A0048.pdf.

Wang, J., J. Huang, L. Zhang, Q. Huang, and S. Rozelle. 2010. "Water Governance and Water Use Efficiency: The Five Principles of WUA Management and Performance in China." *Journal of the American Water Resources Association* 46 (4): 665–85.

Wang, J., Z. Xu, J. Huang, and S. Rozelle. 2005. "Incentives in Water Management Reform: Assessing the Effect on Water Use, Production, and Poverty in the Yellow River Basin." *Environment and Development Economics* 10 (6): 769–99.

Wang, J., Z. Xu, J. Huang, and S. Rozelle. 2006. "Incentives to Managers or Participation of Farmers in China's Irrigation Systems: Which Matters Most for Water Savings, Farmer Income, and Poverty?" *Agricultural Economics* 34 (3): 315–30.

WOCAT (World Overview of Conservation Approaches and Technologies). 2007. *Where the Land Is Greener—Case Studies and Analysis of Soil and Water Conservation Initiatives Worldwide*, edited by H. Liniger and W. Critchley. Wageningen: Technical Centre for Agricultural and Rural Cooperation, United Nations Environment Programme, Food and Agriculture Organisation of the United Nations, and CDE.

World Bank. 2017. *Myanmar's Public Expenditure Review: Fiscal Space for Economic Growth.* Yangon: World Bank Myanmar.

World Bank. 2019. "From Rice to RICE: Integrated Strategies for Myanmar to Realize a More Resilient, Inclusive, Competitive and Environmentally Sensitive Food System." Presentation by Indira Janaki Ekanayake, Mateo Ambrosio-Albala, Steven M. Jaffee. Report AUS0000470, 2.

Zhang, L., N. Heerink, L. Dries, and X. Shi. 2013. "Water Users Associations and Irrigation Water Productivity in Northern China." *Ecological Economics* 95 (November): 128–36. https://doi.org/10.1016/j.ecolecon.2013.08.014.

Zhang, W., F. Jiang, and J. Ou. 2011. "Global Pesticide Consumption and Pollution: With China as a Focus." *Proceedings of the International Academy of Ecology and Environmental Sciences* 1 (2): 125–44.

APPENDIX E

Transformative Innovations Affecting Agri-Food Systems

This appendix presents information on transformative innovations complementary to that in chapter 5.

INNOVATIONS FOSTERING PRODUCTIVITY

E-services to smallholders

Agri-food system efficiency depends on the diffusion of innovations, tools, and knowledge to all participants in the value chain, especially smallholders, who are often slower or less able to adopt new knowledge and institutional innovations. Smallholders are generally less inclined to make "risky" investments in new innovations, which they may not understand how to use in their own context. Information and communication technology (ICT) applications in agricultural extension, weather and price information, and banking have been successfully promoted in several countries and regions since about 2005 (World Bank 2011, 2012); box E.1 offers a few examples from East Asia.

Blockchain technology

Blockchain technology has found innovative uses in several sectors, including agriculture (FAO 2019). The agriculture sector has a number of blockchain applications in the pipeline. Blockchain applications in the agri-food sector feature new opportunities to improve productivity, inclusivity, and ways to monitor compliance with environmental and food safety requirements. Blockchain applications are being pursued to improve transparency and traceability in agricultural supply chains, manage land rights and registrations, offer index insurance based on smart contracts, support cooperative development, address water and sanitation, and enhance financial inclusion as well as digital identity and remittances. Blockchain may also have uses in monitoring, verifying, and reporting on green or climate bonds that were created to fund projects with positive environmental or climate benefits (FAO 2019).

Agricultural insurance. Low-cost agricultural insurance schemes are increasingly viewed as mechanisms for providing social protection to the increasing numbers of people affected by floods or droughts, helping to lessen the impacts they suffer because of such events. However, the rate of adoption of insurance

BOX E.1

Examples of e-service applications reaching smallholders in East Asia

E-extension services. The design and execution of extension services differ between countries in East Asia. Most are public sector and cooperative driven, with an increasing role for the private sector and nongovernmental organizations. Several countries have experimented with e-extension, and examples have been documented elsewhere (World Bank 2011, 2012). Most e-extension services are subnational, nimble, and managed by nonpublic actors. An example of a public-driven comprehensive e-extension comes from China. China developed online agriculture information platforms to implement the 12316 hotline, a specialized agriculture information service system project of the Ministry of Agriculture; the service aims to provide fast and effective dissemination of information on agriculture technology, markets, and policies. Subsequently, the hotline was combined with websites, agriculture TV programs, mobile phone short messages, and multimedia messages to create a new public welfare 12316 Information Service System for Agriculture, Farmers and Rural Areas. The online version gets about 8.6 million visits daily, and the hotline takes more than 20 million phone calls annually. Approximately 80,000 "Farmer Friendly Information Stations" had also been built as of 2017; the aim is to cover 80 percent of China's villages by the end of 2020 (ADB 2018).

Weather services. In the era of climate change, targeted services that give farmers access to high-quality data on temperature, rainfall, wind, and soil moisture will be particularly important. Research shows that pest and disease forecasting services and weather-based crop insurance programs are among the fastest-growing agricultural climate services in the Asia region.

Market information. Services that help farmers connect to markets and consumers more efficiently, or provide real-time pricing information on commodities, also help improve outcomes for farmers. For example, Indonesia has supported the launch of five apps that are meant to support farmers; one of the apps, Pantau Harga, allows users to track food prices across the country.

Integrated services. There is some Asia-based evidence of low-cost tools and collaborations between business and government diffusing knowledge and technology through the supply chain. Information and communication technology tools are delivering information and analytics to farmers.

Vinaphone, a Vietnamese state telecom company, launched a service that turns mobile phones into "farm assistants," providing agricultural information services (such as weather forecasts, plant disease alerts, guidelines on government policies, and advice on plant diseases) through a low-cost monthly package, delivered through social media. Similar services exist in Indonesia, through 8Villages.

Collaborations with governments and larger companies can improve smallholders' exposure to technology and knowledge. Formal collaborations can also help make product supply chains more visible to end-consumers. Indonesia's Partnership for Indonesia Sustainable Agriculture, established in 2012, tightens collaboration between the state, companies, and smallholders to improve the production of commodities including cocoa, dairy, maize, palm oil, potatoes, rice, and soybean.

Another Indonesia-based initiative, the Oil Palm Development Plasma Programme, works to improve farmers' knowledge of oil palm cultivation and management best practices.

Financing services. Access to finance is one of the fundamental components of agricultural extension efforts. Even the best innovations may not percolate through the supply chain if farmers are reluctant to borrow, or if banks refuse to lend. Asia has several promising initiatives in play, mostly in mobile money. USAID's mStar project, operational in Bangladesh, sought to reduce reliance on physical cash transactions among smallholders by rolling out mobile services across the rice value chain. This enabled farmers to receive loans and payments from partners via mobile phones, with USAID using its convening power to work with mobile operators on reducing disbursement fees. There have also

continued

Box E.1, *continued*

been experiments with micro-financing with partners down the supply chain, such as the Integrated Supply Chain Corn Partnerships, co-developed with Syngenta, which brings together banks, retailers, and grain traders and provides micro-financing to farmers, helping them access markets in the Indonesian province of West Nusa Tenggara. The scheme increased farmers' productivity by 10 percent, earning them 1.8 million Indonesian rupiah (approximately $144) more per hectare, in turn giving businesses a steady supply of commodities. Another collaboration, Farmer Nature Net, established savings and credit associations and cooperatives, which improved farmers' liquidity. These innovations need to be complemented by policies removing structural barriers that impede access to finance, such as lack of land rights. Secure land tenure can encourage farmers to invest in improving farm productivity, and also provide collateral for obtaining finance.

Source: Green 2018.

products by the rural poor remains relatively low. The mechanisms that are in place to validate claims and to effect payouts are still time consuming, which is one reason index-based insurance is not chosen as the first risk-mitigation strategy by smallholder farmers. Index insurance based on smart contracts can automate and greatly simplify the process, thereby facilitating instant payouts to the insured in the case of adverse weather incidents.

An example with regard to agricultural insurance is Arbol—Smart Contract Weather Risk Protection for Agriculture, which is an Ethereum platform built to protect farmers, livestock operators, and other agribusiness groups against weather risk via a peer-to-peer funding mechanism. Arbol leverages publicly available weather application programming interface services to automatically evaluate index thresholds and make payments to participants. This offering provides several benefits over traditional crop insurance and index insurance schemes (Jha, Andre, and Jha 2019).

Agricultural supply chains. Blockchain can assist in providing an immutable record from the provenance to the retail store of a product, and thereby bring increased transparency in agricultural supply chains. This can give consumers increased trust in the products they buy, and it is also an opportunity to reward the producers who use good agricultural practices to cultivate their produce. This would eventually lead to sustainable farming practices and responsible consumption.

Land registrations. Blockchain-based applications could provide an incorruptible ledger of land records. Especially in the case of the rural poor, if the record is linked effectively to sovereign or digital identification, the safekeeping of land records even in times of natural disasters or wars would not be an issue.

Green or climate bonds. Blockchain could be used in sustainably monitoring, verifying, and reporting on green or climate bonds. Green bonds were created to fund projects that have positive environmental or climate benefits. With increases in bond value, it is necessary to have effective tracking, traceability, and verification mechanisms to help increase investors' trust in climate-smart initiatives. Carbon credits and trading in them could benefit from the trust that blockchain offers.

Forestry. Hangzhou Yi Shu Blockchain Technology Co., Ltd in China aims to use blockchain for forestry economic development and rural poverty alleviation. The Spanish Ministry of Agriculture, Fisheries and Food also plans to apply blockchain technology to develop the forestry industry. The operating group, ChainWood, aims to improve the traceability and efficiency of the wood supply in Spain by implementing blockchain technology in the industry's logistics. The group is set to develop cloud-based software that would improve the transparency of forestry processes—such as the creation of solid wood, disintegration, cellulose paste, and biomass—by applying blockchain, big data, and machine learning (FAO 2019).

Microbiome technologies to enhance crop resilience

The plant microbiome—the environment of microorganisms in and around the roots, in the soil, on the leaves, and within the plant itself—has the potential to change modern agriculture. When applied directly to the surface of seeds and to plants themselves, microbiome technologies can complement or replace chemical agriculture products. Companies are now trying to identify the organisms that are most beneficial in different environments and to create products based on them. To realize this potential, continued advancement will be needed in the methods for analyzing agro-ecological conditions and in creating products for specific contexts, complemented with an adequate enabling environment (WEF 2018). Indigo's work focuses on beneficial microbes, the bacteria, viruses, and fungi that naturally coexist with plants. Some of these microbes work with plants, helping them overcome typical stresses during the growing season. Indigo's technology screens samples to identify beneficial microbes, which it then develops into a seed coating. Indigo launched its first product, a treatment for cotton plants in 2017. Since then, it has brought to market microbial products for corn, wheat, soybeans, and rice (WEF 2018).

INNOVATIONS IN THE FOOD CHAIN

Solar-powered storage solutions may be an option in off-grid areas or areas with unreliable or expensive electricity. Stand-alone solar may offer new opportunities to make use of new innovations in ICT as well as to support otherwise risky interventions in livestock production, irrigation, and processing. For example, in Uganda, stand-alone solar systems were used in different value chains to energize pumping and irrigation systems, refrigerator units (cooling milk, storing veterinary drugs and fresh produce), milling and other processing equipment, and obviously to maintain connectivity (Economic Consulting Associates Limited 2020).

In recent years, the renewable energy technologies and battery storage systems that power off-grid systems have advanced rapidly, with quality rising and costs declining. Several factors will be critical to realize a significant impact. Renewable energy and battery storage technologies must continue to advance, and their associated costs must continue to decline. For end users and companies, the up-front costs of off-grid clean energy technologies are often larger than fossil fuel alternatives even when there are long-term economic benefits. Pay-as-you-go mechanisms, made feasible by mobile technologies, make renewables more affordable. Because energy systems tend to be expensive, quality-assured products and standards are essential. Additionally, it is important to have a good understanding of the total life-cycle costs of each renewable application because there are cases in which the environmental costs of producing these systems may be greater than the gains from using them on-field. Renewable energy and storage technologies have tremendous potential, especially for many countries in Africa and Asia. If the cost issues can be addressed, both people and the environment will benefit (WEF 2018).

INNOVATIONS IN FOOD CONSUMPTION AND NUTRITION

Growing urbanization and income levels have driven consumer interest in nutrition, including reformulation, functional foods, and alternative food sources. Box E.2 provides complementary (to chapter 5) details on fortification, reformulation, and functional foods development in East Asia. Use of 3D (three-dimensional) printing is featured in box E.3.

Nutrigenetics technology identifies how genetic variations affect people's responses to nutrients. Quick and cost-effective DNA analysis is possible because of advances in the understanding of the human genome as well as in computing and data analytics. The ultimate objective is personalized nutrition, that is, the ability to optimize a person's nutritional intake based on genetic makeup. Using this technology, nutritionists and doctors will be able to individualize health and diet recommendations, potentially improving the health of consumers.

By 2030, nutrigenetics is likely to affect the dietary choices of consumers in higher-income countries, resulting in those consumers living longer and healthier lives. Additionally, the technology could have a material impact on the overweight population. If it was adopted by 10–15 percent of the overweight population in high-income and upper-middle-income countries, it could reduce the number of overweight people by 25–55 million. For nutrigenetics to be scaled up, considerable numbers of nutritionists, doctors, and insurers will have to advocate for it. Some consumers may adopt the technology on their own because they are interested in knowing more about their ability to process nutrients. Most consumers, however, will do so only if their health care professionals encourage them and if their health care insurance plans cover the cost. Additional uptake could be driven by reduced life insurance premiums if consumers follow personalized nutrition programs.

Providing dietary information is only one part of the solution, however. Expansion also depends on consumers being adequately educated on healthy dietary practices and in food selection and preparation. Nutrigenetics is still in the early phases of development. The impact of certain genes on absorbing, transporting, storing, or metabolizing nutrients is not yet sufficiently understood (WEF 2018).

BOX E.2

Food fortification, reformulation of food, and functional foods

Food fortification in East Asia. Today, Indonesia, Nepal, the Philippines, and Vietnam all require wheat flour to be fortified with micronutrients including iron, folic acid, and vitamin B12. Legislation in China promotes voluntary fortification. Some Asian countries also require salt to be fortified with iodine, or vitamin A to be added to cooking oil. Rice is trickier to enrich in nations that operate small, decentralized mills, but both Papua New Guinea and the Philippines have made rice fortification mandatory. Food companies in Asia are also focusing on fortification to address concerns about health and nutrition, and to respond to local tastes, product competition, and government regulation

Reformulation involves the adaptation of consumer products to remove harmful substances, or to find natural replacements for ingredients such as sugar and fat. Nutrient profiling covers a range of related interventions, including promoting healthier consumption through marketing and labeling, and reformulating products to remove harmful ingredients or optimize nutritional balance. Research has shown reductions in trans fats in the United States thanks to reformulation since 2000, and in the United Kingdom, a public-private collaboration reduced the salt content of supermarket products by about 20–30 percent, resulting in an estimated 10 percent reduction in salt intake at the population level. Nutrient profiling activities in Indonesia, the Republic of Korea, Singapore, and Thailand have begun although specific targets for nutrient levels were not yet present (Green 2018)

In the long term, food manufacturers are likely to focus on reformulation to meet demand for healthier alternatives. Asian countries and companies are moving toward best practices and collaborate regionally. Product reformulation collaboration between health bodies and food manufacturers is particularly strong in Singapore, driven by the country's Health Promotion Board. It focuses on reducing high salt levels in processed food such as noodles and bread, as well as more natural foods such as fish cakes (Green 2018).

Gluten-free and casein-free foods. The global gluten-free products market was estimated to be $5 billion in 2015. "Free-from" foods are receiving considerable attention, especially gluten-free and casein-free food products. There are a few reasons for this development; for instance, it has been estimated that about 1 percent of the population suffers from celiac disease (an immune reaction to eating gluten) in Indonesia. There is also a perceived image of "free-from" foods as healthier and premium, and the mass media has raised awareness of this subsegment in many countries in Asia (Ecosperity 2018a).

Specific opportunities include developing food products for individuals with various allergies and investing in research to expand the range of food that everyone can safely consume. For instance, restaurants such as Bliss Restaurant and the Swensen's chain in Singapore are willing to make accommodations to their menus (for example, replacing eggs or peanuts) to help customers manage their allergies.

Many companies have conducted research to replace ingredients causing allergies with safer ones such as pulses, seeds, and grains. For example, GoodMills Innovation collaborated with scientists, grain breeders, and nutritionists to introduce a wheat flour that is consumable by people sensitive to gluten and still produces baked goods with the same texture and taste as those made with traditional wheat.

Functional foods are processed foods containing ingredients that aid specific bodily functions and support nutrient intakes in addition to meeting basic nutritional needs. They are essentially foods that have been heavily fortified and enhanced to provide substantial health benefits. This includes food products containing probiotics, prebiotics, plant stanols, and folic acid.

The demand for functional foods will rise as the aging population increases across the world. Furthermore, as more people are educated on the nutritional benefits of specific food ingredients such as

continued

Box E.2, *continued*

turmeric, there will be new demand and markets for companies to enter. The interest in these products is partly being driven by married couples and households with children. Market research estimates that the market value for functional food was $44 billion in 2015 (Ecosperity 2018b).

Functional foods have also been gaining traction in China, with high consumption of products containing probiotics in Beijing, as well as rapid growth in consumption of traditional Chinese cooling herbs in Guangzhou and tonic herbs in Chengdu. In Japan, brands include Suntory-owned Brand's Essence of Chicken (which is the market leader in broths with more than $600 million in sales and presence in 19 countries), Yakult Honsha, and Meiji Group (Nutraceuticals World 2016).

COUNTRY CAPACITY TO EMBRACE TRANSFORMATIVE INNOVATIONS

This section provides complementary details to the section titled "Readiness of developing East Asian countries to embrace transformative innovations" in chapter 5. Some of the details are from chapters 4, 5, and 6.

Capacity in biotechnology largely correlates with economic development and country views about the use of biotechnology—essentially genetic modification and genetic engineering (GE). Overall, genetically modified (GM) crops have been approved in several countries for commercial production (table D.1 in appendix D), and a wide range of crops and traits are either at the experimental stage or in field trials. China, Malaysia, and Thailand have the highest overall capacity to generate and adopt biotechnology innovations. They have, however, different views on the use of biotechnology; for example, Thailand has not approved GM crops for commercial production, whereas Malaysia imports GM crops but does not carry out research on GM crops or GE. Although China imports several transgenic crops and has invested heavily in the development of its own varieties (box 5.5), it has approved the commercial planting of very few GM crops (Wong and Chan 2016). China has also significantly invested in GE as part of a wider, technology-based push to improve agricultural output, and is about to become the global leader (box 5.6). Indonesia (box 6.5), the Philippines, and Vietnam continue to improve their capacity in biotechnology; for example, both Indonesia and the Philippines have endorsed or begun using GE for crop improvement. They have commercialized several GM crops, and adoption by farmers is widespread. Biotechnology capacity in Cambodia, the Lao People's Democratic Republic, and Myanmar is currently low,[1] but adoption of, for example, crops with enhanced production traits is feasible.

East Asian countries are at different stages in the development and adoption of precision agriculture, but all can adopt the least complex applications. While Japan, the Republic of Korea, Malaysia, and Singapore tend to be the frontrunners in developing and adopting precision agriculture, China (from national to county levels) has also invested heavily in precision agriculture and overall mechanization of agriculture.[2] Both Thailand and Vietnam are highly interested in emerging technologies, the Internet of Things, ICT, and vertical farming

BOX E.3

Three-dimensional printing of food and machinery

Food for the most affluent. Another high-tech innovation is three-dimensional (3D) printing, which can manufacture farming equipment and even food. Additive manufacturing layers ingredients to create bespoke products on demand and is attracting investment from global players including Mondelez and Hershey's. 3D food printers could even become home appliances, allowing the production of everything from pizza to chocolates, and reduce food waste in restaurants by allowing tailored portion sizes and easier adaptation of products to exclude allergens. However, 3D printing is not currently a significant disruptive force for the Asian food system. It is most likely to be a product for higher-income consumers and those visiting high-end restaurants. The wealthy hubs of Asia, such as Hong Kong SAR, China; Seoul; Taipei; and Singapore, have been exposed to visiting roadshows and exhibitions showing off the wonders of 3D food, but it is not likely to be significant for the mass population of Asia.

Machinery. A more important trend is the application of 3D printing to the creation of bespoke agricultural tools and machinery. This is already a growing industry in the United States, and there is evidence of 3D-printed farm technology in low-income contexts such as Myanmar. Farmers seeking specific tools and technologies could find 3D printing useful.

Source: Green 2018.

automation (box 6.6). Indonesia, the Philippines, and Vietnam have also adopted diverse, less advanced, and less costly precision agriculture and field-monitoring tools, such as drones, suited to their needs, yet commercial adoption of precision tools tends to be slow.

Most East Asian countries have tried to optimize the outreach and coverage of extension with diverse e-extension services (discussed in chapter 6). China is the regional leader in promoting ICT in agriculture and has also integrated e-services into its nationwide decentralized extension system (box 6.9, box E.1).

Urban farming is most sophisticated and widespread among the most urbanized countries of the region but is expected to take greater hold more widely as urbanization accelerates. In addition to Malaysia, rapidly urbanizing China is far ahead of its less affluent neighbors in promotion (for example, agro-parks), research, and solutions in urban farming (box 5.7). Thailand has also embraced urban farming, both low- and high-tech. Indonesia, the Philippines, and Vietnam have adopted new innovations in urban farming; for instance, in the Philippines vertical farming is part of the formal agriculture agenda and it has been piloted and tested (along with hydroponics). The rest of the countries engage in urban farming but it typically entails low-tech applications, such as growing kits, information packs, and polytunnel farming, offered by urban agriculture associations and start-ups. These innovations and technical advice may, however, spur urban farming to new levels in the less urbanized follower countries. The common challenges to urban farming include regulatory issues (such as quality control), land availability, access to innovations, and access to extension services, finance, and consumer acceptance (box 5.7). In addition to enabling innovation in technology and adoption (for example, through incentives for the private sector), governments can use a range of levers to encourage urban agriculture, from streamlining regulations (on safety and quality) to providing extension services and funding.

Most East Asian countries are making progress in developing or adopting climate-smart agriculture (CSA) and other sustainable agriculture approaches, but wider adoption faces barriers. Countries with strong R&D systems or functioning decentralized extension services,[3] supported by the private sector and international agricultural research centers, tend to be better off. China, being by far the dominant emitter of agricultural greenhouse gases in the region, has pursued diverse programs to address agricultural pollution and climate change (appendix D). Among its peers, Vietnam is ahead of the curve, and has pursued sustainable practices and CSA (boxes 4.2 and 4.3; table 4.4; appendix D), yet there is more to do given its thin knowledge base in sustainable production. The agriculture sector of the Philippines is quite different and underdeveloped, also reflected in its low uptake of CSA practices.[4] Cambodia, Lao PDR, and Myanmar have limited CSA work but can collaborate, for example, with international agricultural research centers and the regional networks on CSA (appendix F). Myanmar is one of the most vulnerable countries in the region to climate change, yet it is only in the early stages of CSA adoption.[5]

Most innovation on food loss and waste and the circular economy takes place in high-income countries, and, in East Asia, China, Japan, Korea, Malaysia, and Singapore are leading the way. Although the stakeholders in Indonesia, the Philippines, and Vietnam have become conscious of food loss and waste, limited adoption or innovation currently takes place. Regional approaches, for example, on logistics pave the way for the lower-income neighbors to benefit.

All East Asian countries already fortify or have the capacity to fortify foods, yet capacity on more sophisticated nutrition innovations is limited. Capacity or perhaps propensity for reformulation is mostly concentrated in Korea and Singapore as well as in Indonesia and Thailand (Green 2018). In the long term, food manufacturers are likely to focus on reformulation to meet demand for healthier alternatives.[6] Functional foods have created momentum mostly in China and Singapore. Capacity for nutrigenetics is nearly nonexistent across the region. Science, technology, and innovation capacity requirements for meat alternatives vary, ranging from low (plant-based alternatives) to high (lab-grown meat). Today, only Japan has the capacity for lab-grown meat while China has resorted to importing lab-grown meat from Israel.

NOTES

1. Although Myanmar has begun using GE tools.
2. China has, for example, issued a series of policy documents for guidance, implemented numerous lower-tech demonstration projects in rural areas, and experimented with precision agriculture. Examples include a satellite navigation system that combines high-precision positioning technology with sensor technology for accurate monitoring of soil moisture, farm machinery autopilot control, and direct seed precision planting; the intelligent rice bud production system; and real-time data collection through temperature and moisture sensors in greenhouses to achieve intelligent micro-spraying and electric shutter ventilation control (ADB 2018).
3. For instance, some areas of China, Vietnam, plantation crop sectors in Malaysia, Thailand, and Indonesia with well-functioning extension.
4. Low adoption is mostly a result of poor availability of and access to improved seed, insufficient financial resources to cover investment costs, and the limited resources of extension services.

5. Myanmar's government allocated most agricultural public expenditures to rice cultivation, while the sustainable management of livestock and manure treatment received little in the budget. The large investments in irrigation and flood control were not climate smart (World Bank 2017).
6. For instance, about 1 percent of the population suffers from celiac disease (an immune reaction to eating gluten) in Indonesia (Green 2018).

REFERENCES

ADB (Asian Development Bank). 2018. *Internet Plus Agriculture: A New Engine for Rural Economic Growth in the People's Republic of China*. Metro Manila, Philippines: ADB.

Economic Consulting Associates Limited. 2020. "African Solar Designs. Market Assessment Study: Stand-alone Solar for Productive Uses in Uganda." Draft Final Report, February 13, 2020. Submitted to World Bank by Economic Consulting Associates.

Ecosperity. 2018a. "Feeding Urban Asia: New Approaches for Providing Safe, Nutritious, and Affordable Food." Ecosperity, Singapore.

Ecosperity. 2018b. "Prevention Is the Best Medicine: Rethinking Healthcare in Asia." Ecosperity, Singapore.

FAO (Food and Agriculture Organization of the United Nations). 2019. *E-Agriculture in Action. Blockchain for Agriculture. Opportunities and Challenges*. Bangkok: FAO and International Telecommunication Union.

Green, Adam. 2018. "Food 4.0: The Future of Food Innovation in Asia." Economist Intelligence Unit Limited.

Jha, S., B. Andre, and O. Jha. 2019. "ARBOL: Smart Contract Weather Risk Protection for Agriculture." Unpublished.

Nutraceuticals World. 2016. "Key Trends in Functional Foods & Beverages for 2017." http://www.nutraceuticalsworld.com/issues/2016-11/view_features/key-trends-in-functional-foods-beverages-for-2017.

WEF (World Economic Forum). 2018. *Innovation with a Purpose: The Role of Technology Innovation in Accelerating Food Systems Transformation. System Initiative on Shaping the Future of Food Security and Agriculture*. Geneva: World Economic Forum together with McKinsey & Company.

Wong, A. Y.-T., and A. W.-K. Chan. 2016. "Genetically Modified Foods in China and the United States: A Primer of Regulation and Intellectual Property Protection." *Food Science and Human Wellness* 5 (3): 124–40. https://www.sciencedirect.com/science/article/pii/S2213453016300076.

World Bank. 2011. "Information and Communication Technologies for Agriculture e-Sourcebook." World Bank, Washington, DC.

World Bank. 2012. *Agricultural Innovation Systems. Investment Sourcebook*. Washington, DC: World Bank.

World Bank. 2017. *ICT in Agriculture: Connecting Smallholders to Knowledge, Networks, and Institutions*. Washington, DC: World Bank. https://openknowledge.worldbank.org/handle/10986/27526.

APPENDIX F

Country Agricultural Innovation Systems Assessment

The assessment of agricultural innovation systems (AIS), and the key elements of agriculture research and development (R&D), extension, partnerships and co-innovation, skills, governance, and the enabling environment are discussed in chapter 6. The following boxes offer details on foreign private R&D in China, international cooperation in R&D, and public-private partnership models for R&D in Indonesia.

PRIVATE AGRICULTURAL R&D

Private sector agricultural R&D is on the rise in the region. Box F.1 features China's diverse strategies in acquiring foreign private R&D.

INTERNATIONAL COLLABORATION

International collaboration overall, and on R&D, is increasingly important for East Asian countries struggling with new emerging challenges. Box F.2 describes international R&D collaboration in the countries of the Association of Southeast Asian Nations (ASEAN).

Key mechanisms for cross-country cooperation in the region include bilateral and multilateral development agencies, membership in international research networks (box F.3), and cooperation between higher education institutions (HEIs). All three methods of cooperation foster international partnerships with individual countries, the European Union, ASEAN, multilateral organizations, HEIs, and international agricultural research centers. Furthermore, ASEAN is a platform for regional cooperation on innovation (OECD 2017b). China along with others has participated in collaborative arrangements with ASEAN and ASEAN+3 countries (China, Japan, the Republic of Korea) for overseas visits, study-abroad programs, training, and exchange and cooperation projects, which have significant spillover effects on the ASEAN countries.[1] Although China has been active in the integration of international collaboration frameworks in the agri-food area, it has yet to advance substantially in exploiting the potential of these frameworks.

BOX F.1

Foreign private agricultural research and development in China—through acquisition

Despite being the world's largest agricultural producer, China has restricted foreign company participation in seed and other agricultural input markets to minority shares in joint ventures with Chinese firms (Pray and Fuglie 2015). In addition, enforcement of intellectual property laws is seen as relatively weak, approval of genetically modified crop varieties has been limited to cotton, and state-owned companies continue to play a major role in supplying agricultural inputs to Chinese farmers. The spending of 1,305 Chinese agribusiness firms on agricultural R&D was estimated to be $244 million in 2006 (Hu et al. 2011).

Despite low levels of agricultural research and development spending by both foreign and domestic firms in China, Chinese companies have used direct acquisitions of foreign companies to gain access to their capacities and technology, for instance, acquisitions by the state-owned enterprise ChemChina of the France-based specialty feed company Adisseo (for $500 million in 2005) and the Israel-based agricultural chemical company Adama (for $2.4 billion in 2011). In 2011, the privately owned Chinese company Shuanghui International acquired the U.S.-based meat processing company Smithfield Foods for $4.72 billion, including its subsidiary Smithfield Premium Genetics, one of the world's largest privately held pig breeding operations. In the same year the state-owned Chinese machinery manufacturer Sinomach acquired McCormick France Corporation, a French farm machinery parts manufacturer. But by far the most significant potential acquisition is of Syngenta by ChemChina. In February 2016, ChemChina's $43 billion offer was accepted by Syngenta and is currently undergoing regulatory review. Agricultural research and development spending in 2014 by all these foreign acquisitions was $1.49 billion ($1.38 billion by Syngenta alone).

BOX F.2

International cooperation on agricultural research and development

Agricultural research and development (R&D) is a significant and increasing area for international research cooperation in member states of the Association of Southeast Asian Nations (ASEAN). The share of internationally co-authored publications in total publications provides an insight into the degree of international collaboration by ASEAN countries on agricultural R&D. Between 2004 and 2008, about 57 percent of publications by ASEAN countries on food, agriculture, and biotechnology had international co-authorship, exceeding the share of publications in Australia, China, Japan, and the Republic of Korea (UNU-IIST 2011). More recent analysis suggests that the European Union, Japan, and the United States are significant collaborators. Between 2004 and 2013, 16 percent of ASEAN journal publications on agriculture, fisheries, and forestry by Southeast Asia–based authors had at least one co-author from Europe, 10 percent of publications had at least one co-author from Japan, and 9 percent had at least one co-author from the United States (Lampert and Stefan, forthcoming).

The high share of publications with international co-authorship seen in ASEAN member states points to the openness of researchers from the region to collaboration, and the fact that research agencies have developed strong international research links with international research institutes and development agencies. However, it may also reflect the weaker domestic research capacity of some countries; the share of publications with international co-authorship ranged from 38 percent in Malaysia to about 90 percent in Cambodia, the Lao People's Democratic Republic, Myanmar, and Vietnam (UNU-IIST 2011). These numbers suggest that, in some ASEAN countries, international research cooperation is an important mechanism for building research capacity.

continued

Box F.2, *continued*

China's co-authored patents represented 22 percent of all its agri-food patents; a similar share for the Organisation for Economic Co-operation and Development (OECD) was 12 percent, and 17 percent for the European Union. About 10 percent of China's co-authored patents were in the areas of agriculture and food processing, which is below OECD and BRIICS (Brazil, the Russian Federation, India, Indonesia, China, South Africa) averages. Between 2007 and 2012, Chinese authors produced a relatively small number of joint publications with foreign co-authors in agri-food sciences, constituting one of the lowest shares of total agricultural science publications in a country in an international comparison. Less than 15 percent of agriculture science publications by Chinese scholars had foreign co-authors, while almost every second publication in both the OECD and in the European Union, and one-third in the BRIICS had at least one. These indicators suggest that, although China has been active in the integration of international collaboration frameworks in the agri-food area, it has yet to advance substantially in exploiting the potential of these frameworks (OECD 2018a).

Source: OECD 2017b.

BOX F.3

International networks in East Asia

Apart from international collaboration with the CGIAR and multinational organizations (such as the European Union of the United Nations, the Food and Agriculture Organization, the Association of Southeast Asian Nations) and bilateral relationships with many countries, the East Asian countries represented by the main government research agencies are also members of international research networks, including for research on cross-cutting issues. Notable research networks and platforms include the Global Research Alliance on Agricultural Greenhouse Gases (GRA), Conservation Agriculture in Southeast Asia (CANSEA), the Asia-Pacific Association of Agricultural Research Institutions (APAARI), and SEA-EU-NET.

Indonesia, Malaysia (represented by the Malaysian Agricultural Research and Development Institute), the Philippines, Thailand, and Vietnam (represented by the Vietnam Academy of Agricultural Sciences) are part of the GRA, and are variously members of the livestock, paddy rice, croplands, and integrative research groups. Member countries of the GRA collaborate on the research, development, and extension of technologies and practices to help deliver more climate-resilient food systems without increasing greenhouse gas emissions.

CANSEA is a network that promotes the development of innovative farming systems based on an agro-ecological approach that combines rural development with environmental preservation. Its membership includes Cambodia (represented by the Ministry of Agriculture, Forestry and Fisheries), Indonesia (represented by the Indonesian Agency for Agricultural Research and Development), the Lao People's Democratic Republic (represented by the National Agriculture and Forestry Research Institute), Thailand (represented by Kasetsart University), and Vietnam (represented by Ministry of Agriculture and Rural Development agencies); and the network is linked with institutions in Australia, China, and France.

APAARI promotes the development of national agricultural research systems in the Asia Pacific region through interregional and interinstitutional cooperation. The main research agencies from Malaysia (the Malaysian Agricultural Research and Development Institute), the Philippines (Bureau of Agricultural Research and the Philippine Council for Agriculture, Aquatic and Natural Resources Research and Development), Thailand (Department of Agriculture), and Vietnam (Ministry of Agriculture and Rural Development) are regular members of APAARI.

Source: OECD 2017b.

AGRICULTURAL EXTENSION SERVICES

The section in chapter 5 titled "Readiness of developing East Asian countries to embrace transformative innovations" discusses the development of agricultural extension in East Asia. Brokers (box F.4) represent a new type of service provider. Brokers can facilitate increasingly complex multistakeholder innovation processes. Research-extension-farm (-firm) links are another important function in AIS. Box F.5 features organizations supporting research-extension-farm (-firm) links in Vietnam.

BOX F.4

Innovation brokers

Context. The stakeholders in agricultural innovation have become more numerous and their interactions more complex. The creation and fostering of effective coalitions among actors is often hindered by incomplete information about what potential partners can offer and by different incentive systems for public and private actors, differences between indigenous and formal knowledge, social differences that cause exclusion of certain actors, or ideological differences.

The role of the honest broker resembles a broadened notion of the role of a process facilitator (Klerkx and Leeuwis 2009). Any advisory service or related individual or organization can broker, connecting farmers to different service providers and other actors in the agricultural food chain. Examples include research organizations, national and international nongovernmental organizations, specialized consultancy firms, temporary projects, government programs, and farmers' organizations (Klerkx, Hall, and Leeuwis 2010). Although public organizations such as extension services and research institutions could perform innovation brokering as part of their mandates, many often retain a linear, transfer-of-technology mindset and lack the capacity to fulfill this role (Devaux et al. 2009; Rivera and Sulaiman 2009). Innovation brokers can also be independent, specialized organizations with a skill set especially tailored to innovation brokering.

Activities. Brokers may analyze the context and articulate demand, compose networks, and facilitate interaction. Innovation brokers have been found at the supranational (across several countries), national (country), regional (province, district), and (sub)sectoral or commodity level (such as dairy or horticulture), but these levels may also mix (for example, when dealing with cross-cutting value chain innovations).

Benefits. Through investments in innovation brokering, communication between the multiple actors can improve greatly. By providing fresh insights and a mirror for self-reflection, innovation brokers stimulate clients to look beyond their current situation and constraints. Impartial, honest brokers, because of their less-biased position and the overview of the system that they can provide, can forge contacts between parties that would normally not cooperate. They can also mediate more easily in the case of conflict.

Innovation brokering can be expected to have both immediate and long-term results. Direct results are expected through market innovations that arise when producers respond better to the needs of agribusiness and agribusiness operators develop a better understanding of production systems. Brokering can facilitate technical innovation by improving how agricultural research service providers target serious bottlenecks in production or processing or by inducing required institutional change on the part of policy makers and legislators. Over the longer term brokering should improve how the overall innovation system functions.

BOX F.5

Research-extension links and knowledge brokers in Vietnam

Relevant agricultural research results are often not shared with farmers. Despite the significant number of agro-scientists working at the local level, interaction between the various actors is limited. However, the number of Vietnamese organizations acting as a bridge between knowledge generators and users has grown in recent years, but their skills and expertise remain limited. These organizations are mostly public bodies, with the private sector playing a limited role. For example, there are several centers for promoting technology transfer and information under the local Department of Science and Technology and the Centers for Promoting Agricultural Activities under the local agencies for Agriculture and Rural Development. In addition, technology incubators are linked to polytechnic universities and to a growing number of technology parks, which play a role in facilitating knowledge diffusion.

On the local level, the State Agency for Technical Innovation (SATI) is supported by the departments for science and technology in the provinces, which are the main connectors for innovation networks. SATI formed a joint program with the Japanese government that has supported about 1,000 Vietnamese companies in upgrading their manufacturing operations by buying Japanese equipment. Under the new Technology Transfer Fund, there is a planned Centre for Science and Technology Transfer Services to help local enterprises adopt foreign technology and to transfer know-how and technologies from public research institutes and universities. In addition, SATI organizes technical training and capacity-building activities at the local and provincial levels.

Source: OECD and World Bank 2014.

INSTRUMENTS AND EXAMPLES OF CO-INNOVATION

Indonesian agencies are searching for "models of what works," that is, models that are useful in promoting public-private sector partnerships that mobilize science for innovation. See box F.6.

EDUCATION AND SKILLS FOR AGRI-FOOD SECTOR INNOVATION

The role of education and skills development is crucial for agricultural innovation, both as a producer of skilled personnel and as a source of knowledge and research. Table F.1 summarizes the main findings on agricultural education and skills (with data limitations), drawing on the desk review.

The tertiary and technical and vocational education and training (TTVET) capacity differences across the developing East Asian countries are significant. TTVET systems (number and size of agencies, programs, human and other resources) and their quality vary largely along the development status of the country. All countries have HEIs that offer courses and degree programs in agricultural disciplines (table F.1). In most countries, HEIs fall under the ministry of education, which has often not been the most advantageous institutional arrangement for agricultural education (World Bank 2012). In Cambodia, the Lao People's Democratic Republic, and Vietnam, universities and colleges fall under the ministries of agriculture.[2]

The number of HEIs varies between countries (table F.1). China, Indonesia, Malaysia, Thailand, and Vietnam all have several universities and colleges

BOX F.6

Applied Research on Innovation Systems in Agriculture interventions in Indonesia

Applied Research on Innovation Systems in Agriculture (ARISA) is supporting collaborative projects between research institutes and private sector companies to incubate and deliver technology and business solutions appropriate to smallholder farmers in eastern Indonesia. These projects are supported by capacity building and technical assistance tailored to the individual partnerships. ARISA seeks to identify and analyze opportunities and barriers to the expansion of research-private sector partnership that can help translate and deliver ideas and solutions from research to farmers. Interventions are described below.

The *beef intervention* involves developing a profitable and sustainable beef production system in Sumbawa, Nusa Tenggara Barat (NTB). This has been done through improved engagement of cattle farmers with a traders association (PEPEHANI), individual large traders, and a beef processing company (PT Dharma Raya Hutama Jaya). The research institute partner is the University of Mataram. The intervention aims to improve the incomes of approximately 1,100 cattle farmers in West Sumbawa and Sumbawa districts.

The *maize intervention* involves promoting best practices for dual-cropping models using new hybrid maize with pulses (mung bean and ground nut) on drylands in Nusa Tenggara Barat. The partners are PT Syngenta Indonesia, Bank NTB, and the University of Mataram. The intervention seeks to improve the incomes of about 1,100 smallholder farmers in East and North Lombok.

The *cassava intervention* involves developing integrated modified cassava flour chip clusters for improving the welfare of smallholder farmers in the southern part of East Java. The partners are PT Bangkit Cassava Mandiri (PT BCM), the University of Jember, the KEHATI Foundation, plus a range of enterprise cooperatives. This intervention aims to improve the incomes of approximately 2,800 cassava and sheep farmers in the Jember region of East Java.

The *sugar intervention* involves improving market links, the commercialization of agricultural innovations, and an enabling policy environment for sugarcane development in Madura, East Java. The partners are PT Perkebunan Nusantara X and the Indonesian Sugar Research Institute, along with Trunojoyo University. This intervention seeks to improve the incomes of approximately 1,200 farmers in Madura.

The *dairy intervention* involves developing fodder farming business models for smallholder dairy production in East Java. The partners are PT Nestle and the University of Brawijaya. The intervention aims to improve the incomes of approximately 1,200 dairy and fodder farmers in the Malang region of East Java.

The *shallot integrated pest management (IPM) intervention* involves using IPM for shallot production in East Java. It is a joint ARISA-PRISMA intervention. The main partners are PT NuFarm, PT Nasa, PT Solbi, the University of Gadja Mada, and the Plant Protection Agency. The implementation plan also called for CropLife Indonesia to be involved. The intervention aims to improve the incomes of at least 3,000 shallot farmers in East Java through the adoption of IPM.

Source: Hall et al. 2016.

offering agriculture-, food-, water-, and forestry-related education and degrees. The Philippines is home to a large number of HEIs—in 2011, there were 97 state colleges and universities offering degree programs in forestry and other related courses, and at least 106 offering degree programs in agriculture (Cruz et al. 2013). In Lao PDR, the National University of Laos is the only HEI to offer postgraduate programs in agriculture and food and forestry (there are also five agricultural colleges). A lack of resources potentially affects the quality of agricultural education in many countries. Only a small proportion of university lecturers and researchers at HEIs in Cambodia and Vietnam are qualified at the PhD

TABLE F.1 **Education and skills for agricultural innovation in East Asia**

COUNTRY	VOCATIONAL[a]	TERTIARY	FIRMS	FARMERS
China	—	Several high-quality universities carrying out both education and research. University-industry collaboration exists.	Exchange of personnel between industry and academia.	Professional farmer-training program. ICT skills lacking. Aging farmers.
Indonesia	One higher-level vocational school or academy is equivalent to a bachelor's degree. Another higher secondary school in agricultural development offers vocational education at secondary or high school level. Indonesia also has few national, provincial, and local training centers that provide training to diverse actors. Little information is available regarding the scope.	More than 50 universities and polytechnics offer various agriculture-related courses. Bogor Agricultural University is the predominant one. Limited university-industry collaboration exists. Despite improvements, quality of education is an issue. Important to improve cooperation between firms and education institutions to address skills shortages.	Difficulty finding labor with appropriate skills.	ICT skills lacking; overall education and skills limiting adoption of innovations.
Malaysia	Colleges and training institutes offer nondegree programs in the agriculture-food sector.	Nine universities offer programs related to food and agriculture; six are major universities. University-industry collaboration exists. Although significant attention has been paid to education over the past decades, there is a need to reform and improve industry-academia links to better match labor market needs and R&D capacity.	Difficulty finding labor with appropriate skills, including business, finance, and management skills.	Aging farmers.
Philippines	Strengthen technical and vocational training.	A large number of higher-education institutions offer agricultural programs. In 2011, there were 97 state colleges and universities offering forestry and other related courses, and at least 106 offering degree programs.	Difficulty finding labor with appropriate skills.	ICT skills lacking.
Thailand	Four well-established institutions provide formal technical and vocational education training. Shortage of students. Attention to entrepreneurial skills needed.	Several major research universities, such as Kasetsart University, provide high-quality agricultural education, as do more teaching-focused universities. Curriculum update required, for example, on entrepreneurial skills. Difficult to attract students and lecturers. Limited R&D personnel. Technical skills and ability for multidisciplinary research limited. Important to improve cooperation between firms and education institutions to address skills shortages, that is, continue to promote university-industry collaboration.	Difficulty finding labor with appropriate skills.	Smart Farmer e-learning to address limited farmer capacity. Most farmers with primary education only. ICT skills lacking. Average age over 50.
Vietnam	Limited domestic trainee and apprentice practice. Make college-based and TVET training programs more distinctive and more attractive to improve the relevance of their programs to employers' needs, and to improve quality-assurance mechanisms and service delivery.	Ministry of Agriculture and Rural Development supervises four universities with regard to agriculture, water, and forestry. Several other universities provide degree programs in agriculture. Limited number of PhD-holding lecturers and researchers.	Difficulty finding labor with appropriate skills.	ICT skills lacking.

continued

TABLE F.1, *continued*

COUNTRY	VOCATIONAL[a]	TERTIARY	FIRMS	FARMERS
		Education reform needed to address quality and relevance of labor market skills (currently theoretical and rigid). Changes are needed to improve curricula, to stimulate the recruitment and hiring of teachers, and to align academic programs more closely with labor market needs. A pipeline of science, engineering, and technology graduates is needed to respond to sector needs.		

Sources: Original table for this publication. Global Forum for Rural Advisory Services World Wide Extension Study data set, 2016 (https://www.g-fras.org/en/world-wide-extension-study.html); OECD 2017a, 2017b, 2018a; UNCTAD 2018; World Bank 2012.
Note: — = not available; ICT = information and communication technology; R&D = research and development; TVET = technical and vocational education and training.
a. Data mostly not available for vocational schools.

level—about 16 percent in Vietnam—and about 8 percent of full-time faculty at Cambodia's Royal University of Agriculture (RUA) have PhDs (De la Pena and Taruno 2012). RUA is further constrained by a lack of proper equipment and laboratory facilities (James, Gill, and Bates 2013). Both attractiveness and relevance of agricultural education are constraints in East Asia (box F.7). Box F.8 features an example of university-industry links in Thailand.

Vocational training in East Asia: Challenges and opportunities

Meeting the skills demands of industry is perhaps nowhere more important than in the provision of vocational education and training. Both Malaysia and Thailand have instituted programs to address the perceived inferior status (see the section titled "Innovation capacity and skills for long-term sustainability" in chapter 6). In the lower-middle-income countries of the region, significant deficits of skilled technicians and workers with vocational qualifications remain, despite a desperate need for people with such skills in industry. For example, in Cambodia, much technical and vocational education and training is of relatively low quality, while little information is given to secondary students on potential careers and job opportunities for skilled technicians. In Vietnam, no dual vocational education system exists in which students combine theoretical training with practical training at a company. Improving the availability of, access to, and quality of the technical track is as important as upgrading the academic track, yet most countries in the region need to do considerably more in this regard.

The school and further or higher education system is not the sole organizational mechanism for creating required human capital. Businesses are also important creators of human capital for the innovation system and are not simply employers of human resources. For example, many of the skills associated with various kinds of design, engineering, and associated management activities are often acquired in firms. Yet there are serious deficiencies in the training of workers by firms in the region, especially among small and medium enterprises, which limits knowledge upgrading and undermines progress toward activities with higher value added and higher productivity (OECD 2013).

BOX F.7

Skills upgrade needed in Thailand

Thailand faces increasing demands for science, technology, and innovation-related human resources, at both the vocational and higher levels of education. The number of research and development personnel per population, particularly in the private sector, is relatively low. In addition, certain fields and sectors, such as both large and smaller technology sectors that need to expand (water management and railways) suffer from skills shortages, and researchers and lecturers in agriculture are scarce. Other much-needed cross-sectoral skills in innovation and entrepreneurship (for example, to facilitate the development of businesses in rural areas and the emergence of more technology-oriented firms) are also lacking.

Fewer young people are engaged in agriculture than ever before. Most Thai farmers (70–80 percent) are low-income smallholders with no formal education beyond elementary school. As educational levels have increased in rural areas in recent decades, younger generations have become less likely to follow their parents into farming, preferring to seek opportunities in other sectors. Vocational schools and agricultural colleges have seen a decline in new student numbers, and universities find it hard to attract good students to agriculture-related courses.

Source: UNCTAD 2018.

BOX F.8

University-industry links for addressing challenges in the Thai shrimp industry

University-industry links have been crucial in fighting the diseases endemic to the shrimp industry in Thailand. Thai educational institutions, especially Mahidol University, helped reduce losses from viruses by developing DNA diagnostic probes. This effort was undertaken in conjunction with an industry consortium, the Shrimp Culture Research and Development Company, and has become well institutionalized.

In 2001, Mahidol University, supported by BIOTEC, established Centex Shrimp (Center of Excellence for Shrimp Molecular Biology and Biotechnology), a multidisciplinary laboratory that combined research from a number of the university's departments. In 2000, BIOTEC also helped establish the Shrimp Biotechnology Business Unit to commercialize research and development results (for example, virus test kits and diagnostic training). Revenues from this unit are invested in further research by BIOTEC and Centex Shrimp (Brimble and Doner 2007; UNCTAD 2018).

Yet Thai university-industry collaboration has faced challenges from protection and low levels of innovation that result in few private sector efforts to link up with universities, rigid structures and weak incentives in Thai universities that discourage ties with business, and generally fragmented Thai bureaucracy. Underlying these obstacles is inconsistent support for university-industry links by politicians.

Sources: Brimble and Doner 2007; UNCTAD 2018.

Korea: Vocational education in agriculture

Korea has 472 vocational education specialized high schools with 287,772 students (Republic of Korea, Ministry of Education 2016). Approximately 6 percent of these students study at 37 agricultural high schools, some of which are specialized in specific areas such as horticulture, horse racing, cooking, and herbal medicine. Approximately 40 percent of the graduates of these schools are employed after graduation, and half of those obtain agriculture-related jobs. Another 40 percent proceed to HEIs. The share of graduates who become self-employed farmers is about 1 percent. At the tertiary level, Korea has 37 agricultural colleges in four-year universities and 5 agriculture-specialized two- and three-year colleges. Approximately 30,000 students are enrolled in agricultural colleges and universities. The overall employment rate was 59 percent upon graduation; of those graduates, 62 percent got jobs in agriculture, and the rate of becoming a self-managed farmer is about 7 percent. See box 6.13 in chapter 6 for additional details.

The Korea Agency of Education, Promotion and Information Service in Food, Agriculture, Forestry and Fisheries is responsible for delivering education and training policies. One of its major projects is financial support for agricultural high schools and agricultural colleges. In 2015, 19 high schools and 14 colleges were supported in delivering a practical curriculum to aid students in advancing to agriculture-related careers. In 2016, three agricultural high schools were selected and provided approximately $2 million for each to raise practical competencies required in the agricultural field. In those high schools more than 70 percent of the curriculum should be vocational subjects and more than 70 percent of vocational subjects should consist of experimental subjects (OECD 2018b).

China: Innovative farmer training

The Chinese government has also established a system for training and extension of agricultural machinery technology, including trials, demonstrations, and on-farm technical advice. Local governments and nongovernmental organizations set up agricultural machinery leasing centers, harvesting machinery service centers, and agricultural mechanization associations. These entities provide comprehensive services related to the use of various agricultural machines, compensating for gaps in government services. The government has increased public support to the training of farmers and agriculture cooperatives as part of an initiative to foster the new professional farmer. For example, the central government allocated 1.1 billion Chinese yuan ($175.1 million) in 2015 to carry out training programs for large specialized farmers, operators of family farms, leaders of agricultural cooperatives, personnel from agribusinesses and agricultural service providers, as well as returning migrated workers. This program covers 4 provinces, 20 municipalities, and 500 demonstration counties. Other services include training programs for rural practical talents and for rural leaders with university degrees, the "million vocational school students" plan to increase annual enrollment in vocational school to more than 70,000, and the National Top Ten farmers' funded projects.

Several forms of education and training of the new professional farmer are available in China. First, farmers' cooperatives offer training courses, which often meet the practical needs of farmers for understanding technical issues such as rice cultivation technology. Second, some communities offer farming schools, which are normally designed to help farmers who have lost land find new employment in intensive agricultural

TABLE F.2 **Overall literacy rates and education enrollment in East Asia**

COUNTRY	ADULT LITERACY RATE (%)	PRIMARY SCHOOL ENROLLMENT (%)	SECONDARY SCHOOL ENROLLMENT (%)	TERTIARY SCHOOL ENROLLMENT (%)
Cambodia	81	90	38	13
China	95	89	—	51
Indonesia	95	93	79	36
Lao PDR	85	91	60	15
Malaysia	94	100	72	45
Myanmar	76	98	60	16
Philippines	96	94	66	35
Thailand	93	98	77	49
Vietnam	94	98	36	29

Source: Original table for this publication. World Bank Databank, 2020 (https://databank.worldbank.org/home.aspx).
Note: — = not available.

production operations. Third, evening school is organized in many rural areas under a "one village one product" initiative to produce a village specialty product. Training classes are organized at night to teach production and processing technologies such as planting tobacco and hybrid rice cultivation, off-season vegetable production technology, and rice flour processing technology (OECD 2016).

The education level of AIS actors has implications for the functioning of the AIS. Table F.2 provides data on overall literacy rates and education enrollment in East Asia. Although adult literacy rates are high in most countries, the differences between countries in secondary education are wide. Many countries also experience a learning gap. On average, children born in developing East Asia are expected to complete almost 11.2 years of education by age 18. Even in Malaysia, which has the highest per capita income in the region, the learning gap (years of formal school adjusted for learning) is a high 3.1 years (World Bank 2019).

GOVERNANCE OF AIS

The governance of the AIS should ensure that national priorities are coordinated and communicated clearly, progress is monitored, and policy outcomes and impacts are evaluated against the defined objectives. Sections titled "Misalignment between sector-wide vision and innovation capacity, investment, and governance" and "Better resource use and innovation outcomes from stronger governance of AIS" in chapter 6 discuss the role of AIS governance in fostering agricultural innovation and coordination with the wider national innovation system. Drawing on the desk review, table F.3 summarizes AIS governance structures and main coordination and knowledge brokers engaged in AIS.

TABLE F.3 **Coordination and governance of agricultural innovation systems in East Asia**

COUNTRY	NATIONAL INNOVATION SYSTEM GOVERNANCE	AIS NATIONAL GOVERNANCE	KNOWLEDGE BROKERS AND COORDINATORS	ASSESSMENT AND PRIORITIZATION	SUMMARY OF ISSUES
China	The National State Council oversees and plays a key role in the decision-making for major innovation plans, including coordinating related ministries and agencies at the national level. The Ministry of Science and Technology and the State Intellectual Property Office develop R&D innovation policy, including for agriculture. The NIS encompasses multitier governance at state, province, and municipality levels.	The MOA, Ministry of Water Resources, and the State Forestry Administration implement the agricultural R&D policy, together with the attached public R&D institutions. The National Development and Reform Commission and the Ministry of Finance allocate public funding for innovation and the technological upgrading of various economic sectors, while the National Natural Science Foundation of China plays an important role in allocating resources for scientific research.	Cooperatives, industry associations, and alliances serve as a bridge between other actors.	The National Center for Science and Technology Evaluation oversees evaluation of government-sponsored R&D projects.	Improve the leadership and governance structure of the AIS, coordination between government agencies and public research institutes at national and subnational levels, including the strengthening of the evaluation of research outcomes, to avoid duplication of research efforts and investments.[a]
Indonesia	Agriculture innovation is integrated into national innovation frameworks. Other ministries have a say in AIS. Oversight by DOST. STI councils provide policy advice, priority setting, coordination.	MOA performs oversight functions for the AIS. A small number of research centers make coordination possible. Coordination agency for extension exists at every governance level.	Limited interaction between stakeholders. Absence of knowledge brokers other than special ARISA interventions and commodity boards.	The Indonesian Centre for Agricultural Technology Assessment and Development is a technical unit responsible for assessing research findings, with no apparent link to governance.	Enhance interaction between stakeholders through platforms and knowledge brokers.
Malaysia	The NSRC is mandated to ensure Malaysia's investment in science and technology makes the greatest possible contribution to a high-value economy. One of the focus areas in NSRC is agricultural sciences. Agricultural innovation is integrated into national innovation frameworks. Other ministries have a say in AIS. Oversight by DOST. STI councils engage in policy advice, priority setting, coordination.	The MOA and the Ministry of Plantation Industries and Commodities perform oversight functions for the AIS. The states of Sabah and Sarawak have autonomy.	Commodity boards and industry associations serve as brokers and networking opportunities. Some firm-centered funds and programs serve as brokers.	An M&E mechanism and instruments for evaluating improvements in governance are not in place. The Palm Oil Board (with local and international experts) uses priority setting, M&E, review processes, presenting good practice for AIS.	Mainstream M&E process as part of AIS governance. Enhance stakeholder collaboration through platforms and knowledge brokers. Consider a dedicated agency or body for AIS coordination and governance across states and ministries.

continued

TABLE F.3, *continued*

COUNTRY	NATIONAL INNOVATION SYSTEM GOVERNANCE	AIS NATIONAL GOVERNANCE	KNOWLEDGE BROKERS AND COORDINATORS	ASSESSMENT AND PRIORITIZATION	SUMMARY OF ISSUES
Philippines	Agriculture innovation is integrated into national innovation frameworks. Other ministries have a say in AIS. DOST contributes directly through Philippine Council for Agriculture, Aquatic and Natural Resources Research and Development. The Department of Environment contributes to agricultural research.	Extremely complex agricultural innovation system. PCAARRD is the central coordinating body for agricultural research. Universities have their own coordination.	—	—	Establish a coordination body and practice that conveys views of all actors for agricultural STI and ensures collaboration across R&D and higher education institutions.
Thailand	MOA is part of the National STI Policy Committee. No central policy or mechanism for cohesion with R&D (limited synergies). Oversight by DOST. STI councils provide policy advice, priority setting, coordination.	Absence of coordinating body for agriculture R&D. R&D efforts cover many disparate areas with limited opportunities to build synergies and strategically address complex problems. The Ministry of Agriculture and Cooperatives sets priorities for research centers, universities for themselves.	Concentration of STI resources around Bangkok, no proper channels for integrating smallholder views and diffusion. Rarely demand driven. Strong cooperatives.	No consistent M&E and not linked to long-term planning.	Establish coordination body for agricultural R&D and innovation, covering all related institutions. Improve inclusive priority setting across the country.
Vietnam	Agriculture innovation is integrated into national innovation frameworks. Other ministries have a say in AIS.	MOA and related ministries have oversight for AIS, budget allocation, and implementation. Numerous research centers, most under the umbrella of Vietnam Academy of Agricultural Sciences. A large number of innovation-related strategies and promotional programs without any coordination.	Limited contribution from other than research personnel, albeit integration with NIS. Some public agencies and incubators perform a brokering role.	Weak monitoring and evaluation practice. The Institute for Policy and Strategy for Agriculture and Rural Development has some good experience in developing policy frameworks.	Strengthen R&D and innovation coordination across multiple institutions, strategies, and promotional programs on innovation. Establish a practice for inclusive (including industry) priority setting. Strengthen M&E feedback loops.

Sources: Original table for this publication. Hall et al. 2019; OECD 2017b, 2018a.

Note: — = not available; AIS = agricultural innovation system; ARISA = Applied Research on Innovation Systems in Agriculture; DOST = Department of Science and Technology; M&E = monitoring and evaluation; MOA = Ministry of Agriculture; NIS = national innovation system; NSRC = National Science and Research Council; PCAARRD = Philippine Council for Agriculture, Aquatic and Natural Resources Research and Development; R&D = research and development; STI = science, technology, and innovation.

a. China's 13th Five-Year Plan includes institutional reform to strengthen coordination between AIS actors, such as the establishment of an interministerial coordination mechanism, improving the link between ministries and provincial government, and strengthening public and private collaboration.
The reform of the organization and management of innovation includes the establishment of a network model of STI, a national innovation platform, comprehensive human resources development, and an effective M&E mechanism (OECD 2018a).

NOTES

1. By the end of 2014, China had arranged more than 14,000 overseas visits and study-abroad programs for Chinese scholars, students, and short-term training staff, and invited more than 30,000 foreign experts to China. Over more than a decade, under the ASEAN+2 and ASEAN+3 frameworks, China has initiated more than 150 agricultural exchange and cooperation projects and trained more than 1,000 technical and managerial staff in ASEAN countries through courses in agricultural technology. Since 2004, China has trained more than 4,000 agricultural technical and managerial staff from Africa (OECD 2018a).
2. The findings are, however, limited in this review. Relatively little information is available on TVET education, particularly on vocational training.

REFERENCES

Brimble, P., and R. F. Doner. 2007. "University-Industry Linkages and Economic Development: The Case of Thailand." *World Development* 35 (6): 1021036.

Cruz, R. V. O., R. B. Bantayan, L. D. Landicho, and N. C. Bantayan. 2013. "Reformulating Agriculture and Forestry Education in the Philippines: Issues and Concerns." *Journal of Developments in Sustainable Agriculture* 8: 49–62.

De la Pena, F. T., and W. P. Taruno. 2012. "Study on the State of S&T Development in ASEAN." Committee on Science and Technology of Association for Southeast Asian Nations, Taguig City, Philippines.

Devaux, A., D. Horton, C. Velasco, G. Thiele, G. Lopez, T. Bernet, I. Reinoso, and M. Ordinola. 2009. "Collective Action for Market Chain Innovation in the Andes." *Food Policy* 34 (1): 31–38.

Hall, A., L. Williams, J. Tanaya, M. Usmani, K. Esti, H. Nufaisa, and L. Xie. 2016. "ARISA: Innovation Systems Research Status and Options. Applied Research on Innovation Systems in Agriculture (ARISA) in Indonesia." Technical Report. CSIRO, Canberra, Australia.

Hu, R. F., Q. Liang, C. E. Pray, J. K. Huang, and Y. H. Jin. 2011. "Privatization, Public R&D Policy, and Private R&D Investment in China's Agriculture." *Journal of Agricultural and Resource Economics* 36 (2): 416–32. https://www.jstor.org/stable/23243089.

James, D., T. Gill, and R. Bates. 2013. "Cambodia: Desktop Study." Innovation for Agricultural Training and Education, Pennsylvania State University, State College, PA.

Klerkx, L., A. Hall, and C. Leeuwis. 2010. "Strengthening Agricultural Innovation Capacity: Are Innovation Brokers the Answer?" *International Journal of Agricultural Resources, Governance, and Ecology* 8 (5–6): 409–38.

Klerkx, L., and C. Leeuwis. 2009. "The Emergence and Embedding of Innovation Brokers at Different Innovation System Levels: Insights from the Dutch Agricultural Sector." *Technological Forecasting and Social Change* 76 (6): 849–60.

Lampert, D., and P. Stefan. Forthcoming. "Analysis of Co-Publications and Co-Patenting among ASEAN Countries and their Collaboration with the EU." SEA-EU-NET, Bonn and Vienna.

OECD (Organisation for Economic Co-operation and Development). 2013. *Innovation in South East Asia*. Paris: OECD Publishing.

OECD (Organisation for Economic Co-operation and Development). 2016. *OECD Reviews of Innovation Policy: Malaysia 2016*. Paris: OECD Publishing. http://dx.doi.org/10.1787/9789264255340-en.

OECD (Organisation for Economic Co-operation and Development). 2017a. *Agricultural Policies in the Philippines*. Paris: OECD Publishing. http://dx.doiorg/10.1787/9789264269088-en.

OECD (Organisation for Economic Co-operation and Development). 2017b. *Building Food Security and Managing Risk in Southeast Asia*. Paris: OECD Publishing. http://dx.doiorg/10.1787/9789264272392-en.

OECD (Organisation for Economic Co-operation and Development). 2018a. *Innovation, Agricultural Productivity and Sustainability in China*. OECD Food and Agricultural Reviews. Paris: OECD Publishing. https://doi.org/10.1787/9789264085299-en.

OECD (Organisation for Economic Co-operation and Development). 2018b. *Innovation, Agricultural Productivity and Sustainability in Korea*. OECD Food and Agricultural Reviews. Paris: OECD Publishing.

OECD (Organisation for Economic Co-operation and Development) and World Bank. 2014. *Science, Technology and Innovation in Viet Nam*. Paris: OECD Publishing. http://dx.doi.org/10.1787/9789264213500-en.

Pray, C. E., and K. Fuglie. 2015. "Agricultural Research by the Private Sector." *Annual Review of Resource Economics* 7 (1): 399–424. https://doi.org/10.1146/annurev-resource-100814-125115.

Republic of Korea, Ministry of Education. 2016. *Statistical Yearbook of Education*.

Rivera, W., and V. R. Sulaiman. 2009. "Extension: Object of Reform, Engine for Innovation." *Outlook on Agriculture* 38 (3): 267–73.

UNCTAD (United Nations Conference on Trade and Development). 2018. *Science, Technology & Innovation Policy Review. Thailand*. Geneva: UNCTAD.

UNU-IIST (United Nations University International Institute for Software Technology). 2011. *Analyses of Research Strengths of SEA Countries for SEA-EU-NET under Task 4.9 Bibliometric Analysis of S&T Strengths in Southeast Asia (Deliverable # 1)*. UNU-IIST, Macao SAR, China.

World Bank. 2012. *Agricultural Innovation Systems. Investment Sourcebook*. Washington, DC: World Bank.

World Bank. 2019. *Managing Headwinds. East Asia and Pacific Economic Update*. Washington, DC: World Bank. doi:10.1596/978-1-4648-1412-9.

www.ingramcontent.com/pod-product-compliance
Ingram Content Group UK Ltd.
Pitfield, Milton Keynes, MK11 3LW, UK
UKHW050119190725
460951UK00007B/95